STUDIES IN MATHEMATICAL BIOLOGY

PART II: POPULATIONS AND COMMUNITIES

MAA STUDIES IN MATHEMATICS

Published by

THE MATHEMATICAL ASSOCIATION OF AMERICA

Studies in Mathematics

The Mathematical Association of America

Volume 1: STUDIES IN MODERN ANALYSIS
edited by R. C. Buck

Volume 2: STUDIES IN MODERN ALGEBRA
edited by A. A. Albert

Volume 3: STUDIES IN REAL AND COMPLEX ANALYSIS
edited by I. I. Hirschman, Jr.

Volume 4: STUDIES IN GLOBAL GEOMETRY AND ANALYSIS
edited by S. S. Chern

Volume 5: STUDIES IN MODERN TOPOLOGY
edited by P. J. Hilton

Volume 6: STUDIES IN NUMBER THEORY
edited by W. J. LeVeque

Volume 7: STUDIES IN APPLIED MATHEMATICS
edited by A. H. Taub

Volume 8: STUDIES IN MODEL THEORY
edited by M. D. Morley

Volume 9: STUDIES IN ALGEBRAIC LOGIC
edited by Aubert Daigneault

Volume 10: STUDIES IN OPTIMIZATION
edited by G. B. Dantzig and B. C. Eaves

Volume 11: STUDIES IN GRAPH THEORY, PART I
edited by D. R. Fulkerson

Volume 12: STUDIES IN GRAPH THEORY, PART II
edited by D. R. Fulkerson

Volume 13: STUDIES IN HARMONIC ANALYSIS
edited by J. M. Ash

Volume 14: STUDIES IN ORDINARY DIFFERENTIAL EQUATIONS
edited by Jack Hale

Volume 15: STUDIES IN MATHEMATICAL BIOLOGY, PART I
edited by S. A. Levin

Volume 16: STUDIES IN MATHEMATICAL BIOLOGY, PART II
edited by S. A. Levin

Studies in Mathematics

Volume 16

STUDIES IN MATHEMATICAL BIOLOGY

PART II: POPULATIONS AND COMMUNITIES

S. A. Levin, editor
Cornell University

Published and distributed by
The Mathematical Association of America

Library of Congress Catalog Card Number 78-53425

Complete Set ISBN 0-88385-100-8
Vol. 16 ISBN 0-88385-116-4

Printed in the United States of America

Current printing (last digit):

10 9 8 7 6 5 4 3 2 1

ACKNOWLEDGEMENTS

Professor Guckenheimer's contribution, "Isochrons and Phaseless Sets," is reprinted with permission of Springer-Verlag, New York, from the *Journal of Mathematical Biology*, 1 (1975), pages 259-273; Professor Kauffman's paper, "The Large Scale Structure and Dynamics of Gene Control Circuits: An Ensemble Approach," is reprinted from the *Journal of Theoretical Biology*, 44 (1974), pages 167–190, by permission of Academic Press, New York; Appendix 3 of Professor May's article, "Mathematical Aspects of the Dynamics of Animal Populations," is a reprint of "Niche Overlap as a Function of Environmental Variability" by Robert M. May and Robert H. MacArthur, which first appeared in *The Proceedings of the National Academy of Sciences U.S.A.*, 69 (1972), pages 1109–1113; the figures in Professor Oster's article, "The Dynamics of Nonlinear Models with Age Structure," are reprinted with permission of Springer-Verlag, New York, from "Dynamics of density dependent populations models", an article which appeared in the *Journal of Mathematical Biology,* Volume 4, No. 2, and was co-authored by Professor Oster with Professors Guckenheimer and Ipaktchi; a portion (pages 158–168) of Professor Segel's contribution, "Mathematical Models for Cellular Behavior," is reprinted from "Mathematics Applied to Deterministic Problems in the Natural Sciences" (pages 22–30) by Lee Segel and C. C. Lin, with permission of Macmillan Publishing Co., Inc., New York; Professor Winfree's article, "Patterns of Phase Compromise in Biological Cycles," is reprinted from the *Journal of Mathematical Biology*, 1 (1974), pages 73–95, with permission of Springer-Verlag, New York; Professor Zeeman's paper, "Primary and Secondary Waves in Developmental Biology," is reprinted with permission of the publisher, American Mathematical Society, from *Lectures on Mathematics in the Life Sciences*, Copyright © 1974, Volume VII, pages 69–96.

INTRODUCTION

PART II: POPULATIONS AND COMMUNITIES

The mathematical theory of biological populations has a rich and long history dating back over half a century and including the fundamental work of Fisher, Volterra, Lotka, Kostitzin, Haldane, and Wright. For those interested in the historical literature, two excellent references exist: *The Golden Age of Theoretical Ecology*, edited by F. Scudo and J. Ziegler, is a fine collection of the major contributions of Volterra, Lotka, Kostitzin, and Kolmogorov; *The Origins of Theoretical Population Genetics*, by W. Provine, traces the history of early population genetics theory and provides an entrée to that literature.

This volume provides a comprehensive overview of the state of modern population theory and its implications for community structure. The static aspects of the theory of communities, basically statistical in nature, have occupied a large portion of the literature, but are not discussed here; neither is numerical taxonomy nor ecosystem modelling. Those interested in such matters are referred to the extensive bibliography at the close of this section.

The central biological principle of this volume is natural selection which, when coupled with interspecific ecological pressures, determines which genotypes shall survive and thereby shapes individual populations and communities. The mathematical foundation lies in the theory of population genetics (treated in the final three chapters) and in the theory of the dynamics of single populations and of several interacting populations. In the opening paper, Robert May develops the deterministic mathematical theory of homogeneous populations, building from the single population to the level of the ensemble of many interacting populations; Donald Ludwig follows May's paper and complements it with an exploration of the effects of the introduction of stochastic effects, which also play a role in the reprinted contribution by May and MacArthur.

The theory of population dynamics as developed especially in May's paper is basically dynamical systems theory, and this has provided much of the basic fare in theoretical population biology since Volterra. A cornerstone of population and community ecology, the competitive exclusion principle (variously modified, amended, enshrined, denied, and villified over the years), is essentially a consequence of such mathematical treatment. The dominant theoretical approach to community dynamics has been by way of systems of ordinary differential equations of the form

$$\frac{dx_i}{dt} = F_i(x_1,\ldots,x_n; y_1,\ldots,y_m), i=1,\ldots,n, \tag{1}$$

in which x_i denotes the density of species i and y_j the level of some environmental parameter. If specifically

$$F_i = x_i f_i(R_1(x_1,\ldots,x_n),\ldots,R_p(x_1,\ldots,x_n)), \tag{2}$$

in which $R_1,\ldots,R_p$ are resource levels and the functions are continuous for $x_1,\ldots,x_n \geqslant 0$, then it is the case that

1. The per capita rate of growth of any species is bounded on every compact subset of the positive orthant. Moreover, $F_i = 0$ if $x_i = 0$; this excludes fixed immigration rates into the region of interest.

2. The resource levels "track" species densities; that is, each R is completely determined by the vector of x's. Basically, this may be interpreted as saying that the resources adjust on a faster time scale than the other species densities.

This is a fairly restricted system, yet subsumes a large portion of conventional population models. Indeed the system defined by (2) is more general than it may first appear. For example, other resources than $R_1,\ldots,R_p$ may be considered. Any resource that does not adjust on the fast time scale may simply be treated as an x variable. Correspondingly, the definition of "resource" need not be restrictive. Any "fast" variable may be treated as an R variable. Traditional competitive systems, such as the "Lotka-Volterra" competition equations

$$\begin{aligned}\frac{dx_1}{dt}&=r_1x_1(K_1-x_1-\alpha x_2)/K_1,\\ \frac{dx_2}{dt}&=r_2x_2(K_2-x_2-\beta x_1)/K_2,\end{aligned}\tag{3}$$

a discussion of which can be found in many of the references (such as Slobodkin, 1961) at the end of this section, are of the basic type prescribed by (2) with resource availabilities taken to be simple linear decreasing functions of species densities. Further, the classical predator prey equations, such as systems (17) or (18) of May's paper, may be interpreted as being of this form, but with three trophic levels (the lowest of which adjusts on a fast time scale).

The generic predator-prey system is of the form

$$\begin{aligned}\frac{dE}{dt}&=A(E,V),\\ \frac{dV}{dt}&=B(E,V),\end{aligned}\tag{4}$$

in which E stands for predator (or exploiter) level, and V for prey (or victim). V is, of course, interpretable as a resource, but one which may adjust "slowly." On the other hand, if the prey adjusts on a "fast time scale," then in general (with suitable constraints on the function B) V will reach an "equilibrium" level $V(E)$ which tracks changes in E. Then, on the slower time scale

$$\frac{dE}{dt} = A(E, V(E)); \tag{5}$$

and this form includes such single population models as the logistic (equation (3) of May's paper). Hence the logistic may be viewed as a degenerate case of a predator-prey system in which the prey equilibrates instantaneously.

Building on the system (1)–(2), Volterra and others stated (in various forms and by different names) the principle of *competitive exclusion**, which in its simplest form states that in a stable community, no two species can share the same limiting resource. Mathematically, this derives from the fact that a system such as (1)–(2) cannot possess an asymptotically stable equilibrium for $n > p$ (R. McGehee and R. Armstrong, 1977, *Journal of Differential Equations*, 23:30–52, provide the most complete discussion). Thus the principle can be directly extended to any number of species and resources, and the notion of resource replaced by the more general notion of *limiting factor* (Levin, 1970, *Am. Nat.* 104:413–423; Whittaker and Levin, 1975, pp. 30–34).

It is easy to find ecological situations that make the principle inapplicable (because the hypotheses are not met). Species may thwart it by temporal variation and temporal partitioning, by spatial displacement, and by adjustments in genetics, age structure, or other demographic parameters. Moreover, the concept of a limiting factor is one of the most slippery imaginable, and the question of how distinct such factors must be to be considered independent is a very difficult one. The reprint of May and MacArthur is an example of an especially sanguine attempt to confront this problem, but the issue remains less than completely resolved. Nonetheless, despite its weaknesses, the exclusion principle remains one of the most useful organizing tools for community research, and is the one theoretical concept around which perhaps the majority of field studies are organized. Because what we see in

*This terminology seems to be due to W. L. Brown, Jr., 1958, *Am. Sci.* 46:151.

nature is most likely the result of long evolution and competitive displacement, it is difficult to document competition taking place in the field; rather what we observe is the result of long periods of adaptation to avoid competition. Thus the competitive exclusion principle in its various forms is difficult to test; but when accepted as a deductive consequence of its assumptions, it is exceptionally powerful in forcing the question "if species are managing to coexist, how do they do so"? This is what makes the notion of competitive exclusion such an important part of the theoretical arsenal of both armchair and field ecologists.

May and MacArthur in their paper address another question: the determinants of species diversity; under particular assumptions, they establish that for species of fixed niche width environmental unpredictability alone sets one upper bound to diversity. (Evolution may increase diversity by narrowing species distributional patterns along gradients, but that is a separate question.) On the other hand, MacArthur and Wilson (1967), concerned primarily with insular situations, view the steady-state number of species as determined by a balance between immigration rates and extinction rates, and thus far short of the upper limit implied by May and MacArthur. In deriving their result, May and MacArthur rely on the fact that near the upper limit, the rate of species extinction rises sharply with number of species. Thus, were immigration rates high enough to place the immigration-extinction balance near that upper limit, the equilibrium (*sensu* MacArthur and Wilson) would be relatively insensitive to immigration rate and in fact of the order of the upper bound itself. The community would be numerically saturated, species turnover rates high, and the individual species relatively transient. In such an environment, with extinction being a common event, population distributional adjustments to one another would be expected to be less finely attuned than in less saturated environments, in which species distributions along gradients should represent tradeoffs between intraspecific and interspecific pressures. Both the insular and saturated ecological communities represent conceivable alternatives, and hence the two approaches to species diversity reasonable extremes. The relative importance of each is the key question, and this depends to large extent upon the scale of inquiry; as is discussed in the paper by Levin, many non-insular environments are most appropriately viewed as mosaics of islands. But whatever determines the local limits to diversity on an ecological time scale, on the longer evolutionary scale the situation seems different. Here not only the temporal but also the spatial scale is larger, and hence a saturation theory seems less justified; there are too many possibilities for evolutionary innovation. An evolutionary theory of ecological displacement is mandated.

The introduction of population structure into theory is of essential importance, both to approach the evolutionary questions and to understand better the ways in which species do coexist. Basically this is the focus of the rest of the volume, including analysis of age structure (Rubinow, Oster), spatial structure (Levin), and genetic structure (Ewens, Karlin, Nagylaki). The general theory of age-structured populations is treated by George Oster, and includes both discrete and continuous time models. It is the former which have been most popular in ecology until recently. The development of continuous time models arose primarily in a somewhat different setting, that of proliferating cellular populations such as are important in microbial organisms and tumor growth; Sol Rubinow precedes Oster's paper with an introduction to that continuous theory. Levin's treatment of populations in spatially heterogeneous environments also makes use at a different level of the theory developed by Rubinow, utilizing it to examine environmental stability. The reappearance of the same basic model in three different settings (Rubinow, Oster and Levin) demonstrates that the same techniques are often valuable in approaching very different biological problems, a point made in the introduction to Volume 15.

Understanding the evolutionary questions raised in the first part of the volume emphasizes the need for an effective theory of population genetics, and the last three papers provide a very current view of the state of the theory. Ewens develops the classical approach and then confronts one of the most disputative of current problems, the importance of deterministic versus random factors in evolution. Karlin provides what is probably the most comprehensive of all theoretical discussions of the complications of multiple loci. Finally, Nagylaki discusses the genetics of spatially distributed populations, a problem of great importance for evaluating the tradeoffs between local variation in selection pressures and the "mixing" effects of migration. There are still a wealth of other new directions, including consideration of random environments (see for example S. Karlin and U. Lieberman, *J. Math. Biol.* 2(1) 1975) and the beginnings of an integration of theoretical genetics and ecology (see for example S. Levin and J. D. Udovic, 1977, *Am. Nat.*, 111:657–675; S. Levin, 1978, in P. F. Brussard (ed.), *Genetics and Ecology: The Interface*, Springer-Verlag, Heidelberg; B. Charlesworth, 1976, in American Mathematical Society Lectures on Mathematics in the Life Sciences, Volume 8). Only when such integration is attained will the understanding of the evolutionary questions be complete. There is still much to be done.

I wish to express my gratitude to the Mathematical Association of America for commissioning this two-volume set, and to the authors for their consistently excellent contributions. I hope that it will serve its purpose as an introduction to the rich possibilities in the study of the mathematics of biology, and am proud to have been associated with it.

SIMON A. LEVIN

READINGS IN THE MATHEMATICS OF POPULATIONS AND COMMUNITIES

N. T. J. Bailey, *The Mathematical Theory of Epidemics*, Griffin, London, 1957.

M. S. Bartlett, *Stochastic Population Models in Ecology and Epidemiology,* Methuen, London, 1960.

M. S. Bartlett and R. W. Hiorns (eds.), *The Mathematical Theory of the Dynamics of Biological Populations,* Academic Press, New York, 1973.

J. F. Crow and M. Kimura, *An Introduction to Population Genetics Theory,* Harper and Row, New York, 1970.

J. M. Cushing, *Integrodifferential equations and delay models in population dynamics,* Lecture Notes in Biomathematics 20, Springer-Verlag, Heidelberg, 1977.

J. M. Diamond and M. L. Cody (eds.), *Ecology of Species and Communities,* Harvard University Press, Cambridge, Mass., 1975.

J. M. Emlen, *Ecology: An Evolutionary Approach,* Addison-Wesley, Reading, Mass., 1973.

J. M. Ewens, *Population Genetics,* Methuen, London, 1969.

R. A. Fisher, *The Genetical Theory of Natural Selection,* Dover, New York, 1958.

G. F. Gause, *The Struggle for Existence,* Williams and Wilkins, Baltimore, 1934, (reprinted 1971 by Dover, New York).

L. R. Gimelfarb, L. R. Ginzburg, R. A. Poluektov, Yu A. Pykh, and V. Ratner, *Dynamical Theory of Biological Populations* (in Russian) Nauka, Moscow, 1974.

N. S. Goel, S. C. Maitra, and E. W. Montroll, *On the Volterra and Other Nonlinear Models of Interacting Populations,* Academic Press, New York, 1971.

A. Jacquard, *The Genetic Structure of Populations*, Biomathematics 5, Springer-Verlag, Heidelberg, 1974. (Translation of French version, published 1970 by Masson et Cie, Paris.)

N. Jardine and R. Sibson, *Mathematical Taxonomy,* Wiley, New York, 1971.

J. N. R. Jeffers (ed.), *Mathematical Models in Ecology,* Blackwell, Oxford, 1972.

N. Keyfitz, *Introduction to the Mathematics of Population,* Addison-Wesley, Reading, Mass., 1968.

M. Kimura and T. Ohta, *Theoretical Aspects of Population Genetics,* Princeton University Press, Princeton, 1971.

K. Kojima (ed.), *Mathematical Topics in Population Genetics*, Biomathematics 1, Springer-Verlag, Heidelberg, 1970.

V. A. Kostitzin, *Biologie Mathématique*, Librarie Armand, Colin, Paris, 1937.

S. A. Levin (ed.), *Ecosystem analysis and prediction*, SIAM, 1975.

R. Levins, *Evolution in Changing Environments*, Princeton University Press, Princeton, 1968.

E. Lewis, *Network Models in Population Biology*, Biomathematics 7, Springer-Verlag, Heidelberg, 1977.

R. Lewontin, *The Genetic Basis of Evolutionary Change*, Columbia University Press, New York, 1974.

A. J. Lotka, *Elements of Mathematical Biology,* Dover, New York, 1956. (Originally published 1925 as *Elements of Physical Biology,* Williams and Wilkins, Baltimore, Md.)

D. Ludwig, *Stochastic Population Theories,* Lecture Notes in Biomathematics, vol. 3, Springer-Verlag, Heidelberg, 1974.

R. H. MacArthur, *Geographical Ecology,* Harper and Row, New York, 1972.

R. H. MacArthur and E. O. Wilson, *The Theory of Island Biogeography,* Princeton University Press, Princeton, 1967.

G. Malécot, *Les Mathématiques de l'Hérédité,* Masson et Cie, Paris, 1948. (English translation published 1969 by W. H. Freeman, San Francisco.)

R. M. May, *Stability and Complexity in Model Ecosystems,* 2nd ed., Princeton University Press, Princeton, 1974.

R. M. May (ed.), *Theoretical Ecology: Principles and Applications,* Blackwell, Oxford, 1976.

J. Maynard Smith, *Mathematical Ideas in Biology,* Cambridge University Press, Cambridge, England, 1968.

———, *Models in Ecology,* Cambridge University Press, Cambridge, England, 1974.

T. Nagylaki, *Selection in one- and two-locus systems,* Lecture Notes in Biomathematics 15, Springer-Verlag, Heidelberg, 1977.

E. P. Odum, *Fundamentals of Ecology,* 3rd ed., Saunders, Philadelphia, 1971.

G. P. Patil, E. C. Pielou, and W. E. Waters (eds.), *Statistical Ecology,* vols. 1–3, Pennsylvania State University, University Park, 1971. (Vol. 1: Spatial Patterns and Statistical Distributions; vol. 2: Sampling and Modeling Biological Populations and Population Dynamics; vol. 3: Many Species Populations, Ecosystems, and Systems Analysis.)

B. C. Patten (ed.), *Systems Analysis and Simulation in Ecology* I, Academic Press, New York, 1971.

———, *Systems Analysis and Simulation in Ecology* II, Academic Press, New York, 1973.

———, *Systems Analysis and Simulation in Ecology* III, Academic Press, New York, 1975.

———, *Systems Analysis and Simulation in Ecology* IV, Academic Press, New York, 1977.

E. C. Pielou, *An Introduction to Mathematical Ecology,* 2nd ed., Wiley-Interscience, New York, 1977.

———, *Population and Community Ecology: Principles and Methods,* Gordon and Breach, New York, 1974.

———, *Ecological Diversity,* Wiley-Interscience, New York, 1975.

R. W. Poole, *An Introduction to Quantitative Ecology,* McGraw-Hill, New York, 1974.

W. Provine, *The Origins of Theoretical Population Genetics,* University of Chicago Press, Chicago, 1971.

F. Scudo and J. Ziegler, *The Golden Age of Theoretical Ecology*, Lecture Notes in Biomathematics, Springer-Verlag, Heidelberg, 1978.

L. B. Slobodkin, *Growth and Regulation of Animal Populations*, Holt, Rinehart and Winston, New York, 1961.

D. P. Smith and N. Keyfitz, *Mathematical Demography*, Biomathematics 6, Springer-Verlag, Heidelberg, 1978.

P. H. A. Sneath and R. R. Sokal, *Numerical Taxonomy, The Principles and Practice of Numerical Classification,* 1973.

V. Volterra, *Leçons sur la Théorie Mathématique de la Lutte pour la Vie,* Gauthier-Villars, Paris, 1931.

P. Waltman, *Deterministic Threshold Models in the Theory of Epidemics,* Lecture Notes in Biomathematics 1, Springer-Verlag, Heidelberg, 1974.

K. E. F. Watt, *Ecology and Resource Management,* McGraw-Hill, New York, 1968.

R. H. Whittaker, *Communities and Ecosystems,* 2nd ed., Macmillan, New York, 1975.

R. H. Whittaker and S. A. Levin, *Niche: Theory and Application, Benchmark Papers in Ecology* 3, Dowden, Hutchinson, and Ross, Stroudsburg, PA, 1975.

G. M. Woodwell and H. H. Smith (eds.), *Diversity and Stability in Ecological Systems,* Brookhaven Symposium in Biology 22, 1969.

CONTENTS

Pages 1–315 refer to Volume 15: Studies in Mathematical Biology, Part I, *Cellular Behavior and the Development of Pattern;* pages 317–624 refer to Volume 16: Studies in Mathematical Biology, Part II, *Populations and Communities.*

MATHEMATICAL ASPECTS OF THE DYNAMICS OF ANIMAL POPULATIONS*

Robert M. May

1. INTRODUCTION

"Though there is endless room for observation in the field of nature, which is boundless, yet investigation (where a man endeavours to be sure of his facts) can make but slow progress; and all that one could collect in many years would go into a very narrow compass."

(Gilbert White, 1789, p. 129)

One aim of theoretical ecology is to provide a framework, within which Gilbert White's facts can be encompassed in an orderly manner, and which points the way for further observation and investigation. Such a framework will rest largely on models which sacrifice detailed realism in an attempt to capture and describe some of the essential dynamical features of interacting populations of plants and animals. The present article outlines some simple differential and difference equations that do this.

The models discussed here are of limited scope. First, none contain any explicit account of evolution; we explore the dynamical properties of various population models, but do not ask how

*This work was supported in part by the National Science Foundation (Grant DEB75-10464). This chapter was written in May 1975.

natural selection shapes the structure of the equations and their parameter values. (A general discussion of the evolutionary significance of population parameters is given, e.g., by Pianka (1974) and Southwood *et al.*, (1974), and detailed mathematical studies have been initiated, e.g., by Roughgarden (1972), Slatkin (1970) and Rocklin and Oster (1975).) Second, the models are, in the main, spatially homogeneous. But spatial heterogeneity is likely to be a central feature in many, if not most, real world situations. (A broad survey of some of the many consequences of spatial heterogeneity has been given, e.g., by May (1974a), and rigorous mathematical treatment of particular sorts of spatial variability are due to Levin and Paine (1974), Levin (this volume), Skellam (1951), Comins and Blatt (1974), Hassell and May (1974).) Third, in the models discussed here populations are regarded as continuous, rather than discrete, variables. This is likely to be sensible so long as all populations are large. However, once a population is likely to attain low values, a demographically stochastic treatment is necessary if one is to give a proper account of extinction probabilities (e.g., Bartlett, 1973; May, 1975a, ch. 2). Fourth, the parameters in most of our models are deterministic. But the real world is unpredictable, and so the environmental parameters (birth rates, carrying capacities, etc.) should incorporate a degree of stochasticity: appendix III gives an example where deterministic and stochastic models lead to qualitatively different conclusions.

Despite these strictures, there are interesting things to be learned from simple deterministic models. In what follows, we study models for single populations, for pairs of populations and for whole communities of interacting populations. Although most of the material is review, some of it is new (e.g., Figs. 2, 5, 6), and some may be cast in an unfamiliar way.

2. SINGLE POPULATIONS

Many isolated laboratory populations, carefully maintained in a controlled environment, may realistically be modeled by a single equation. In the natural world, populations will tend to interact both with their food supply (below them in the trophic ladder) and with their predators (above them in the trophic chain), as well as to depend on various aspects of their physical environment. Even so, it is often useful to regard these biological and physical interactions as passive parameters in an equation for the single population, summarizing them in some overall "intrinsic growth rate", "carrying capacity", or the like.

(i) Continuous growth. If population growth is a continuous process, we are led to study differential equations relating the change in the population, dN/dt, to the population value at any time, $N(t)$.

The simplest such model has a constant per capita growth rate, r, which is independent of the population density:

$$dN/dt = rN. \tag{1}$$

This leads to unbounded exponential growth if $r > 0$, and decrease if $r < 0$.

Setting aside the wishful fantasies of certain religious and political groups, such unbounded growth is not to be found in nature. A simple model which captures the essential features of a finite environment is the logistic equation:

$$dN/dt = rN(1 - N/K). \tag{2}$$

Here the effective growth rate is $r(1 - N/K)$: this is positive if $N < K$, negative if $N > K$, and leads to a globally stable equilibrium point at $N^* = K$. K may be thought of as the carrying capacity of the environment, as determined by food, space, predators, or other things. One way of justifying equation (2) is to regard it as the first term in the Taylor series expansion of more general density dependent models. A plethora of other such models, taken from the ecological literature, is catalogued by May (1975a, ch. 4): all share with equation (2) the essential property of a stable equilibrium point, $N^* = K$, with any disturbance tending to fade away in an exponentially damped manner.

In equation (2), the density dependent regulatory mechanism, as represented by the factor $(1 - N/K)$, operates instantaneously. In most real life situations these regulatory effects are likely to operate with some built-in time lag, whose characteristic magnitude may be denoted by T. A rough way of incorporating such time delays is to rewrite equation (2) as

$$dN/dt = rN[1 - N(t - T)/K]. \tag{3}$$

This delay-differential equation was first introduced into ecology by Hutchinson (1948) and Wangersky and Cunningham (1957), and by now it enjoys an extensive literature (as reviewed, e.g., by May, 1975a, ch. 4). In particular, Auslander, Oster and Huffaker (1974) have shown how equation (3) can arise as an approximation to a fully age-structured description of a single population. More realistically, the regulatory term is likely to depend not on the population at a time exactly T earlier, but rather on some smooth average over past populations (see, e.g., May, 1973):

$$dN/dt = rN\left[1 - \int_{-\infty}^{t} Q(t - t')N(t')dt'\right]. \tag{4}$$

Equation (3) is the special case when Q is a δ-function, $Q(t) = \delta(t - T)/K$.

Whether we use the general equation (4) or the particular equation (3), the qualitative nature of the population dynamics follows from precepts familiar to engineers. If the time delay in the feedback mechanism (namely, T) is long compared to the natural response time of the system (namely, $1/r$), there will be a tendency to overcompensation. For modest values of the time delay this overcompensation produces an oscillatory, instead of an exponential, return to the equilibrium at $N^* = K$; as the time delay becomes longer (as rT exceeds some number of order unity), there is a Hopf bifurcation, and the stable point gives way to stable limit cycles. Specifically, equation (3) has an exponentially damped stable point if $rT < e^{-1}$, and an oscillatorily damped stable point if $e^{-1} < rT < \frac{1}{2}\pi$. For $rT > \frac{1}{2}\pi$, the population exhibits stable limit cycles, the period and amplitude of which are as indicated in Table 1. Appendix 1 sketches the derivation of these and allied results, emphasizing their generic character.

TABLE 1
Properties of limit cycle solutions of equation (3)

rT	$N(\max)/N(\min)$	Cycle period$/T$
1.57	1.00	–
1.6	2.56	4.03
1.7	5.76	4.09
1.8	11.6	4.18
1.9	22.2	4.29
2.0	42.3	4.40
2.1	84.1	4.54
2.2	178	4.71
2.3	408	4.90
2.4	1,040	5.11
2.5	2,930	5.36

Equation (3) or (4) constitutes a minimally realistic model for a single population, in which the density dependent regulatory effects (derived from food supply limitations, or crowding, or whatever) operate with a time delay. The consequent population dynamics can be exponential damping to an equilibrium point, or damped oscillations, or sustained patterns of stable cycles. A wealth of population data can be surveyed in this light. In particular, Krebs (1972), May *et al.*, (1974), Pielou (1974) and Beddington and May (1975) have reviewed sets of data, garnered from field and laboratory, where populations appear to show exponential damping at one temperature, and oscillations (either damped or sustained) at another temperature: this may plausibly be associated with changes in rT produced by the interplay between external temperature and metabolic rates. More specifically, May (1975a, pp. 100–102) has suggested that Nicholson's celebrated blowfly data may have its essential features explained by the simple equation (3); the one-parameter fit illustrated in Fig. 1 assumes the egg-to-adult development time is 9 days, in contrast to the actual figure of 11 days.

Once stable limit cycles arise in either equation (3) or equation (4), their period is roughly equal to $4T$. A qualitative explanation of this fact is as follows: In the first phase of the cycle, the population continues to grow ($dN/dt > 0$) until the earlier population value in the time-delayed regulatory factor attains the

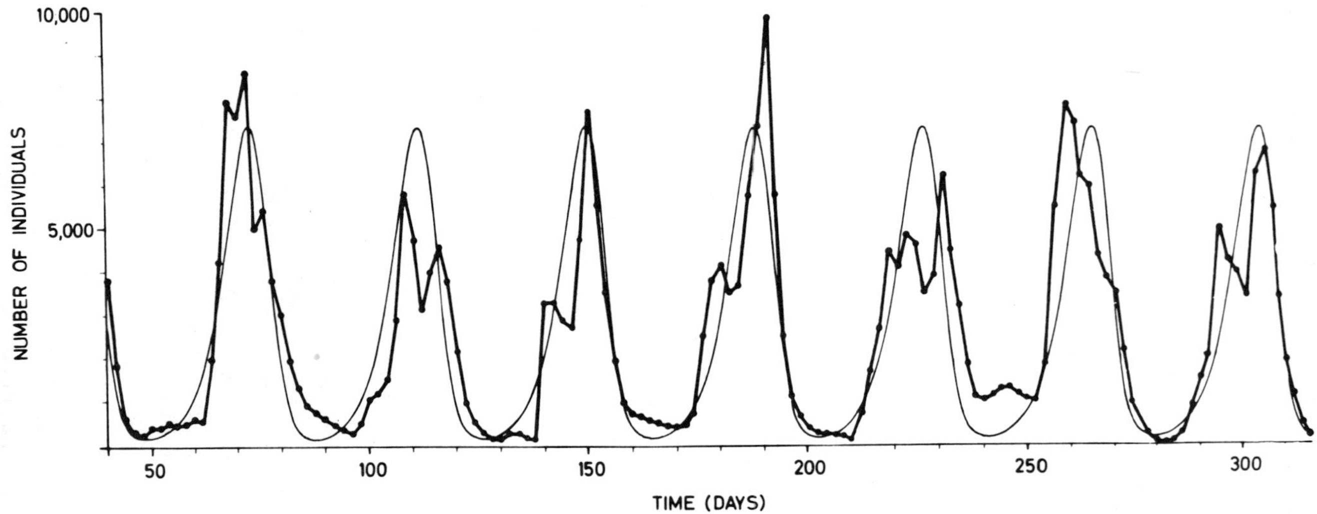

FIG. 1. From the one-parameter family of limit cycles generated by the time delayed logistic equation (3), we display that which best fits the oscillations in Nicholson's blowfly populations (namely $rT = 2.1$: see Table 1). After May, 1975a.

potential equilibrium value ($N(t - T) = K$); at this point, population growth ceases ($dN/dt = 0$), and the population begins an accelerating decline from its peak value. Thus the first phase, where the population grows from around K to the cycle maximum, takes a time T. Similar arguments applied to the subsequent phases of the cycle suggest an overall period of roughly $4T$. The exact results in Table 1 show that this rough rule remains true, even as the amplitude of the cycle (population maximum/population minimum) increases over several orders of magnitude.

The natural world exhibits many instances of 3- to 4-year population cycles, particularly among small mammals in boreal regions. These have excited much attention, from the time of Elton (1942) onward (see, e.g., Chitty, 1950; Pitelka, 1967; Dajoz, 1974). Without attempting to explain the biological mechanisms responsible for density dependent regulation in such populations, it is plausible to assume that in strongly seasonal environments such mechanisms operate with built-in time lags of slightly less than, or of the order of, one year. For beasts with a relatively large intrinsic capacity for population growth, r (so that rT significantly exceeds unity), the upshot will tend to be stable limit cycles, with periods of the order of 4 years. Here is a detail-independent explanation of "wildlife's 4-year cycle". Fig. 2 shows Shelford's (1943) data for a population of the almost legendary lemming, along with a theoretical curve obtained from equation (3) with $rT = 2.4$ (chosen from Table 1 to fit Shelford's numerical data on the amplitude of the cycle) and $T = 9$ months (very roughly the time from the end of one summer to the beginning of the next). Equation (3) is, of course, a very crude model. In particular, it necessarily gives uniformly spaced cycles with a period intermediate between 3 and 4 years, whereas the seasonal northerly environment compels the real cycles to be lock-stepped to an integral number of years, which is sometimes 3, sometimes 4 years.

(ii) Discrete growth. There are many situations where population growth is a discrete process, and where the appropriate models are difference equations relating the population in generation $t + 1$, N_{t+1}, to that in generation t, N_t. Biological examples are provided by many temperate zone arthropod populations, with one short-lived adult generation each year.

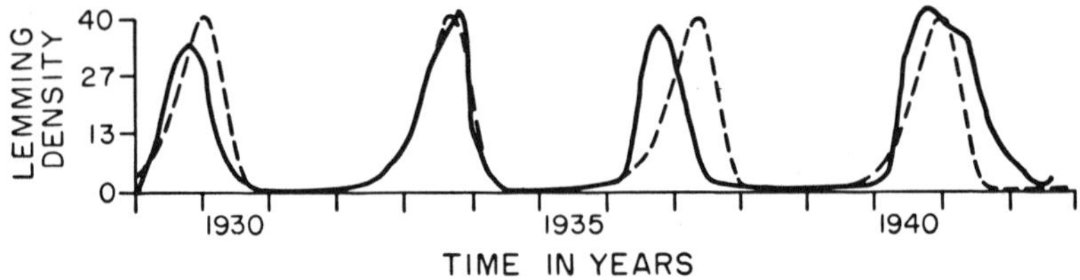

FIG. 2. Shelford's (1943) data on the lemming population in the Churchill area in Canada (expressed as numbers of individuals per hectare), compared with a naive theoretical curve (dashed line) obtained from the simple time delayed logistic equation (3); the time delay T is taken to be a little under one year ($T = 0.72$ yr.). For further details, see the text.

The difference equation analogue of equation (1) is the simple linear equation

$$N_{t+1} = \lambda N_t. \tag{5}$$

Here we have unbounded exponential growth for $\lambda > 1$, exponential decline if $0 < \lambda < 1$.

Among the many density dependent forms which have been proposed as discrete analogues of the logistic equation (2) are the quadratically nonlinear equation

$$N_{t+1} = N_t[1 + r(1 - N_t/K)], \tag{6}$$

and the equation

$$N_{t+1} = N_t \exp[r(1 - N_t/K)]. \tag{7}$$

The biological provenance of these and other similar nonlinear difference equations is catalogued by May and Oster (1975). Using an age-structured model, a similar equation (eq. 13) is derived in Oster's article in this study. Their essential features are a propensity to population growth at low densities, to population decrease at high densities, and a parameter (in this instance, r) which measures the severity of this nonlinear response.

Unlike equations (3) and (4), equations (6), (7) and their friends and relatives do not contain any explicit time lags. However, there is a time delay implicit in the structure of the equation, in the one generation time step between the expression of the density dependent regulatory effects in generation t, and their manifestation in the census data in generation $t + 1$. As above, we then expect exponential damping, or oscillatory damping, or sustained oscillations, depending on the ratio between the time delay ("$T = 1$") and the natural response time ($1/r$). This analogy between delay-differential equations and ordinary difference equations is developed in more detail by May *et al.*, (1974). It sheds light on the behavior of equations (6) and (7) (namely an exponentially damped stable point for $1 > r > 0$, an oscillatorily damped stable point for $2 > r > 1$, and sustained but bounded oscillations for $r > 2$), but gives no hint of the bewildering richness of the spectrum of dynamical behavior which has recently been uncovered in the oscillatory regime ($r > 2$).

A full account of this spectrum of behavior for nonlinear difference equations has been given by May and Oster (1975, see also May,1974b), whose work emphasizes the generic character of the phenomenon. What follows below, and in appendix II, is a brief summary, with numerical examples drawn from equation (7). In his article in this study, Oster says more about this behavior.

What happens when the fixed point (at $N^* = K$) in equation (6) or (7) becomes unstable, as it does for $r > 2$? The argument outlined in general fashion in appendix II shows that it bifurcates, to give two locally stable fixed points of period 2, between which the population oscillates stably in a 2–point cycle. With increasing r, these 2 points in turn become unstable, and bifurcate to give 4 locally stable fixed points of period 4. In this way there arises, by successive bifurcations, an infinite hierarchy of stable cycles of period 2^n, as illustrated in Fig. 3. All else apart, this is of mathematical interest as a particularly simple example of a multiple bifurcation process; most other examples are much more recondite.

This sequence of stable cycles with period 2^n converges on a limiting value of r, r_C say, beyond which the system enters a regime which has aptly been termed "chaotic" (Li and Yorke, 1975a). For any value of r in this domain there are an infinite

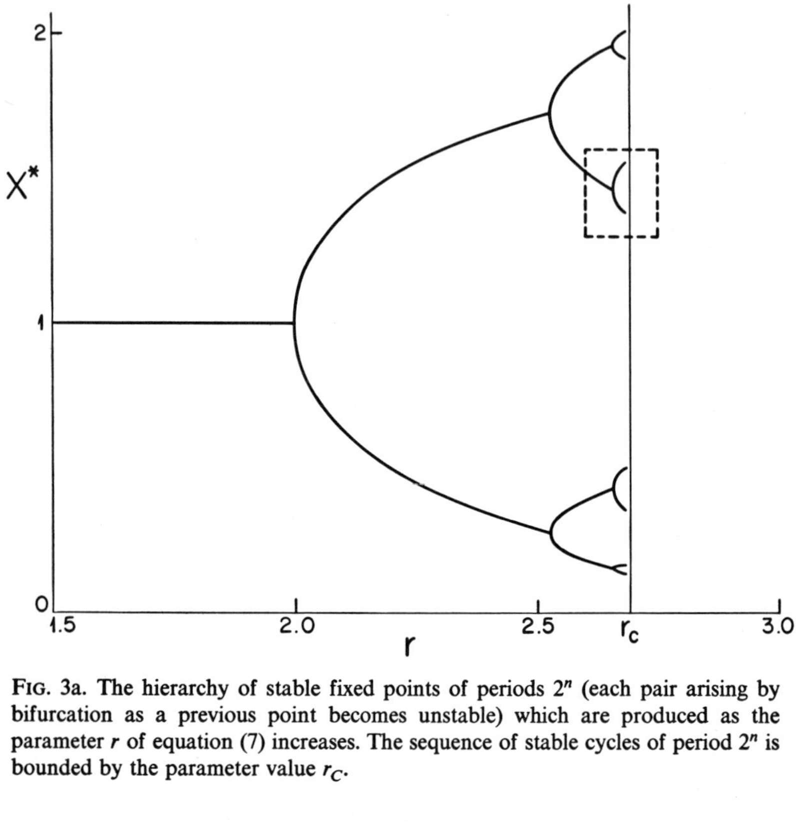

FIG. 3a. The hierarchy of stable fixed points of periods 2^n (each pair arising by bifurcation as a previous point becomes unstable) which are produced as the parameter r of equation (7) increases. The sequence of stable cycles of period 2^n is bounded by the parameter value r_C.

number of different periodic orbits, as well as an uncountable number of initial points for which the population trajectory, although bounded, does not settle into any cycle. In particular, Li and Yorke have proved the surprising theorem that, for *any* first order difference equation, once there exists a 3-point cycle (which happens for $r > 2.828$ in equation (6) and $r > 3.102$ in equation (7)) then there necessarily also exist cycles of every integer period, along with an uncountable number of asymptotically aperiodic trajectories. Li and Yorke (1975a) give a set-theoretic proof; May and Oster sketch a more geometric proof. Another way of characterizing the chaotic regime is to observe that arbitrarily close

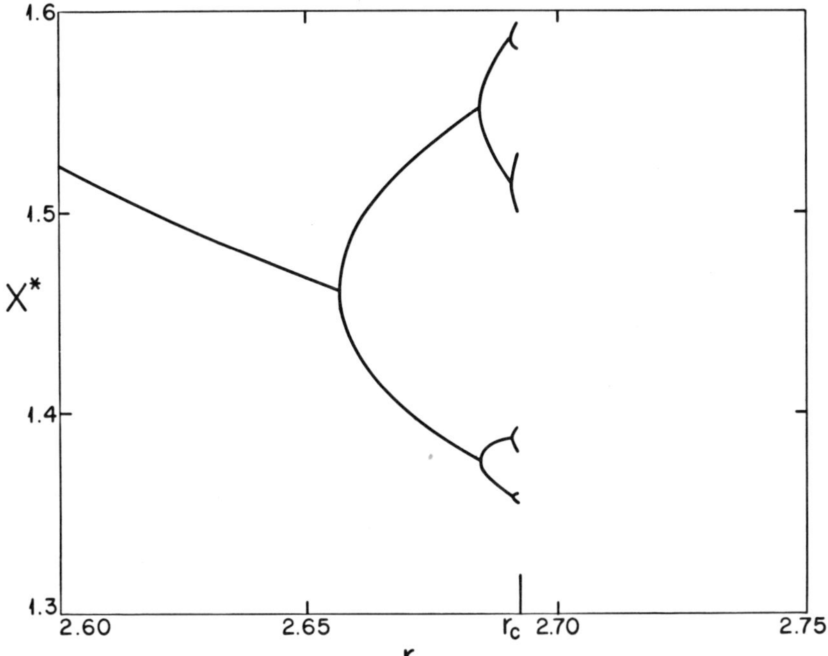

FIG. 3b. A detailed blow-up of the bifurcation process, corresponding to that portion of Fig. 3a within the indicated box.

initial conditions can lead, after sufficient time, to subsequent population trajectories that diverge widely.

Fig. 4 aims to illustrate this range of dynamical behavior for various values of r in equation (7), and Table 2 summarizes the various regimes for both equation (6) and equation (7).

Once in the chaotic regime, the population dynamics of these deterministic models is best described in probabilistic terms. This task has been carried out for equation (7), where it is seen that for largish r the trajectories are almost periodic, with an approximate period $[\exp(r-1)]/r$ (May, 1975b). The general task of finding "invariant measures" for the trajectories in the chaotic region has received attention from Li and Yorke (1975a, b) and others, but

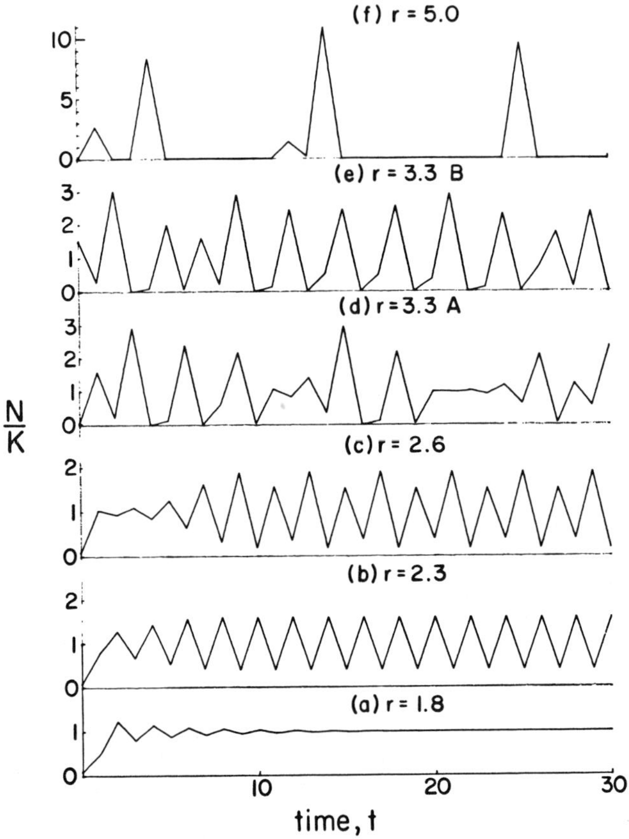

FIG. 4(a to f). The spectrum of dynamical behavior of the population density, N_t/K, as a function of time, t, as described by the difference equation (7) for various values of r. Specifically: (a) $r = 1.8$, stable equilibrium point; (b) $r = 2.3$, stable 2-point cycle; (c) $r = 2.6$, stable 4-point cycle; (d to f) in the chaotic regime, where the detailed character of the solution depends on the initial population value, with (d) $r = 3.3$ ($N_0/K = 0.075$), (e) $r = 3.3$ ($N_0/K = 1.5$), (f) $r = 5.0$ ($N_0/K = 0.02$).

TABLE 2

Dynamics of a population described by the difference equations (6) or (7).

Dynamical behavior	Value of the growth rate, r	
	Equation (6)	Equation (7)
stable equilibrium point	$2.000 > r > 0$	$2.000 > r > 0$
stable cycles of period 2^n		
2–point cycle	$2.449 > r > 2.000$	$2.526 > r > 2.000$
4–point cycle	$2.544 > r > 2.449$	$2.656 > r > 2.526$
8–point cycle	$2.564 > r > 2.544$	$2.685 > r > 2.656$
16, 32, 64, etc.	$2.570 > r > 2.564$	$2.692 > r > 2.685$
chaotic behavior [Cycles of arbitrary period, or aperiodic behavior, depending on initial condition]	$r > 2.570$	$r > 2.692$

much remains to be done before this is in a form where it is a useful tool in the analysis of population data.

Quite apart from their intrinsic mathematical interest, the above results raise awkward biological questions. They show that simple and fully deterministic models, in which all the biological parameters are exactly known, can nonetheless (if the nonlinearities are sufficiently severe) lead to population dynamics which are in effect indistinguishable from the sample function of a random process. Apparently chaotic population fluctuations need not necessarily be due to random environmental fluctuations, or sampling errors, but may reflect the workings of some deterministic, but strongly density dependent, population model. Hassell, Lawton and May (1975) have reviewed a large amount of population data in this light, and have tentatively concluded that, at least for single species situations, population parameters in the chaotic regime are rarely met in nature. This falls far short of settling the issue, partly because there may be a tendency toward bias in the data pool which was surveyed (better studies on populations which are relatively stable), and partly because multispecies systems may enter the chaotic regime more easily than single species ones (a point touched on below, and discussed extensively by May and Oster).

As an anticlimactic postscript to this section, we glance at models for discrete growth in which the density dependent regulatory factors themselves incorporate some explicit time delay T. The result is a delay-difference equation of the general form

$$N_{t+1} = N_t f(N_{t-T}). \tag{8}$$

This includes the models (5)–(7) in the special case $T = 0$. The methods outlined in appendix I may be extended to give a linearized stability analysis of this general equation. Possible equilibrium points, N^*, follow from the equation $1 = f(N^*)$, and the local stability of these fixed points devolves on the slope of $f(N)$ at N^*, i.e., on

$$b \equiv -(N\, df/dN)_{N=N^*}. \tag{9}$$

It may be shown (Levin and May 1976) that N^* is locally stable iff

$$2 \cos(\pi T/2T + 1) > b > 0, \tag{10}$$

and that the region of exponential damping is divided from that of oscillatory damping by the boundary

$$b = T^T/(1 + T)^{1+T}. \tag{11}$$

For both equation (6) and equation (7) it is easily seen that $b = r$ and $T = 0$, so that equation (10) reduces to $2 > r > 0$, and the boundary (11) to $r = 1$: this provides retrospective justification for some of the results asserted above.

(iii) Time-dependent environmental parameters. In the above models, the environmental parameters r and K were taken to be constants. We now examine some of the effects of letting r and K vary in time, in either periodic or stochastic fashion.

There is no difficulty in solving the logistic differential equation (2) for arbitrarily time-dependent $r(t)$ and $K(t)$. By writing $y(t) =$

$1/N(t)$ the equation may be converted to linear form, whence one gets

$$1/N(t) = [1/N(0)] \exp\left[-\int_0^t r(t')dt'\right] + \int_0^t [r(t')/K(t')] \exp\left[-\int_{t'}^t r(t'')dt''\right]dt'. \tag{12}$$

In particular, if r is constant, this has the asymptotic solution

$$N(t) \to \left[r\int_0^t [1/K(t')] \exp[-r(t-t')]dt'\right]^{-1}. \tag{13}$$

These results have been obtained by Poluektov (1974) and by Kiester and Barakat (1974), both of whom use them as a basis for discussing random environmental variability.

Consider the case when r is constant and the carrying capacity $K(t)$ varies periodically:

$$K(t) = K_0 + K_1 \cos(2\pi t/\tau). \tag{14}$$

Here we require $K_0 > K_1$ if $K(t)$ is not to go nonsensically negative during its cycle. The behavior of the population, $N(t)$, will now depend on whether its characteristic response time $(1/r)$ is long or short compared to the periodicity (τ) in the environment. For $r\tau \ll 1$, the solution of equation (13) with equation (14) may be shown to be approximately

$$N(t) \to \left[\frac{1}{\tau}\int_0^\tau dt'/K(t')\right]^{-1}[1 + O(r\tau)]. \tag{15a}$$

That is, neglecting terms of relative order $r\tau$,

$$N(t) \to \sqrt{K_0^2 - K_1^2} = \text{constant}. \tag{15b}$$

In this limit the population averages out the environmental variations: note, however, that this average population value is not simply the average value of $K(t)$, namely K_0, but is less than it. Conversely for $r\tau \gg 1$ we have approximately

$$N(t) \to K_0 + K_1 \cos(2\pi t/\tau)\big[1 + O(1/r\tau)\big], \qquad (16)$$

and the population "tracks" the variations. These tendencies for relatively small r, and for relatively large r, are illustrated by Figs. 5a and 5b, respectively.

Similar considerations apply to situations where $K(t)$ has a more general time dependence, with stochastic components. A detailed exploration of related questions (with spatial instead of

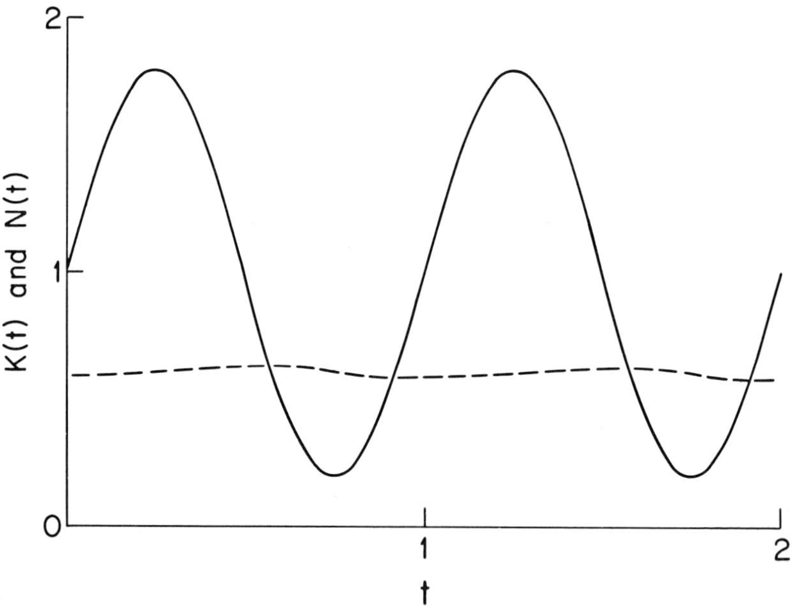

FIG. 5a. Illustrating the way a population ($N(t)$, dashed line) varies if it obeys a logistic equation with a time dependent carrying capacity ($K(t)$, solid line). This figure is for the case when the population's natural response time ($1/r$) is long compared to the periodicity in the environment (τ), i.e., for relatively small r. Specifically, the figure is for equation (2) with equation (14), and $r = 0.2$, $\tau = 1$, $K_0 = 1$, $K_1 = 0.8$; i.e., $r\tau = 0.2$.

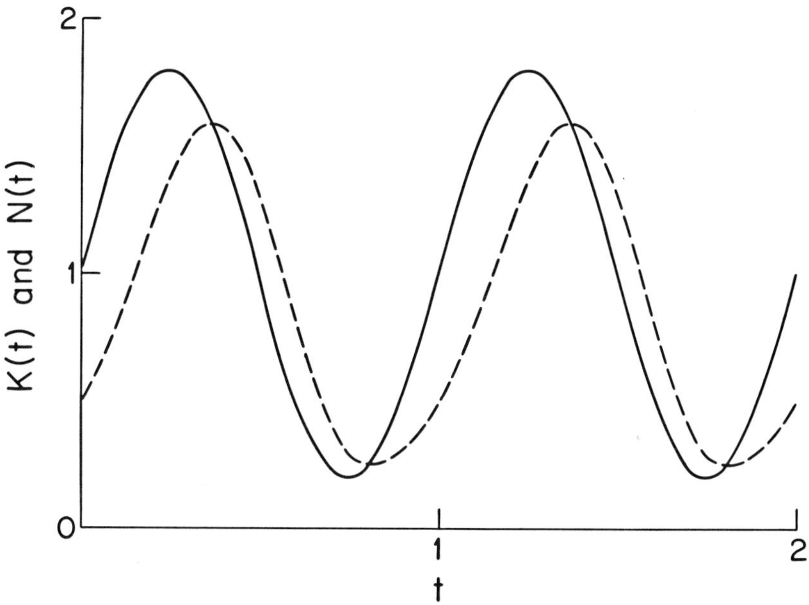

FIG. 5b. As for Figure 5a, but in the opposite limit when the natural response time is short (relatively large r) so that the population tracks environmental variations. The details are as for Figure 5a, except that $r = 10$, whence $r\tau = 10$.

temporal variability) is due to Roughgarden (1974). Reechoing the themes which pervade this article, we find that a population with relatively large r (short response time) is condemned to track environmental fluctuations, whereas those with relatively small r may average over essentially all fluctuations (i.e., over all components of the noise spectrum except those of low frequency, $\omega \lesssim r$).

An investigation of the many other aspects of the way a population behaves in a stochastic environment is largely beyond the scope of the present article. Studies of the relation between the magnitude of the environmental fluctuations and the probability for the population to become extinct were initiated by Lewontin and Cohen (1969) and Levins (1969), and some recent developments are summarized in appendix III; more detailed surveys are due to May (1974a; 1975a, ch. 5) and Ludwig (1974).

3. TWO POPULATIONS: PREY-PREDATOR

(i) Continuous growth. Most ecology texts contain an account of the simplest model for a prey-predator system, namely the classical Lotka-Volterra differential equations

$$dN_1/dt = a\,N_1 - \alpha\,N_1N_2, \tag{17a}$$

$$dN_2/dt = -b\,N_2 + \beta\,N_1N_2. \tag{17b}$$

Here the prey population, $N_1(t)$, has a propensity for unbounded exponential growth (aN_1), which is limited by predators (as measured by the predators' "functional response", $\alpha\,N_1N_2$). The predator population, $N_2(t)$, has an intrinsic death rate ($-bN_2$), and a growth rate or "numerical response" which depends on the prey abundance ($\beta\,N_1N_2$). This system has structurally unstable dynamical properties, namely the pathological neutral stability of the frictionless pendulum; one way of demonstrating this is to note that $V = (\beta\,N_1 + \alpha\,N_2 - b \ln N_1 - a \ln N_2)$ is a conserved quantity. Once perturbed from its equilibrium point, the system oscillates with a period determined by the parameters of the model ($T \cong 2\pi/\sqrt{ab}$), but with an amplitude determined forever after by the initial conditions.

More realistic models may be obtained by modifying the predator-free prey growth term (aN_1) to include density dependence, and by modifying the crude "binary collision" functional and numerical responses ($\alpha\,N_1N_2$ and $\beta\,N_1N_2$) to allow for saturation in the predators' capacity to respond to increasing prey densities. The former modifications tend to stabilize the system, and the latter to destabilize it, as compared with the razor's edge of neutral stability manifested by equation (17). A catalogue of various specific modifications, leading to specific prey-predator models, has been culled from the ecological literature by May (1975a, ch. 4).

The Poincaré-Bendixson theorem may be used to show that essentially all such 2-dimensional prey-predator models have either a stable point or a stable limit cycle. This result was obtained in general form by Kolmogorov (1936), and has been more specifically applied to most of the models in the literature by May (1972; 1975a, ch. 4). An exceptionally nice exposition is in Hirsch and Smale (1974, ch. 12). The system tends to be in tension between the stabilizing prey density dependence, and the destabilizing predator functional and numerical responses: as one or other of the relevant parameters is tuned, the pair of eigenvalues which characterize the stability of the system can be seen to move from the left half of the complex plane (damped oscillations), to cross the imaginary axis into the right half plane, giving rise by a Hopf bifurcation to a stable limit cycle. The neutral stability of equation (17) is reflected in the eigenvalues always lying exactly on the imaginary axis.

If the environmental carrying capacity for the prey, K, is much larger than the equilibrium prey density in the presence of predators, the stabilizing elements contributed to the dynamics by the prey density dependence will be relatively weak. This underlies the "paradox of enrichment", whereby increasing K makes for lowered stability, and eventually for stable limit cycle behavior. The "paradox" was noted (and christened) by Rosenzweig (1971) on the basis of numerical studies, and is possibly exemplified by the larch bud moth in Switzerland (whose population cycles in the optimal part of its range, but is steady in marginal habitats: Baltenzweiler, 1971), and by recent laboratory studies. Stability studies of prey-predator models (Tanner, 1975; May, 1975a, pp. 93, 192–3) have further shown a propensity to stable cycles when the intrinsic growth rate of the prey population exceeds that of its predators. The empirical data reviewed by Tanner (1975) for eight avian and mammalian prey-predator systems accords with these notions: the systems which tended to cycle were indeed those where the prey population had a relatively high growth rate in an environment with a relatively large carrying capacity.

Plant-herbivore systems are, of course, special cases of the general prey-predator relationship, with the plants as "prey" and the herbivores as "predators". Caughley (1975) has recently reviewed the dynamics of ungulate populations from this viewpoint. From the shopping list of qualitatively similar models which are available, he chooses

$$dN_1/dt = a_1N_1[1 - N_1/K] - c_1N_2[1 - \exp(-d_1N_1)], \quad (18a)$$

$$dN_2/dt = N_2[-a_2 + c_2(1 - \exp[-d_2N_1])]. \quad (18b)$$

Here N_2 is the herbivore population density, and N_1 the biomass of the vegetation which sustains it. This system bears out the general remarks made above. It has a stable point if

$$\frac{x(2 \exp x - 2 - x)}{(\exp x - 1 - x)} > d_1K, \quad (19)$$

where $x = (d_1/d_2) \ln (c_2/c_2 - a_2)$; and stable limit cycle behavior otherwise. Note that equation (19) is more easily satisfied for relatively small K, giving an explicit example of the "paradox of enrichment". Caughley assigns plausible values to the parameter values in equation (18) for white-tailed deer in a mosaic of grassland and forest, reindeer on St. Matthew Island, elephant populations in Zambia, and obtains qualitative agreement between the model dynamics and the observed behavior of the real population. He suggests that the patterns of damped oscillations, or of stable cycles, seen in these and other situations may be understood on the basis of such prey (plant)-predator (herbivore) models.

The relation between these 2-species models and the earlier single species models deserves mention. Suppose we try to construct a single species equation for just the prey population in a prey-predator model: the population growth rate will depend on the predator population, which itself is derived (in a formal sense) as some integral over past prey populations. The upshot will be an equation involving only the prey population, but where the density dependent regulatory mechanism depends on past values of the prey population. To be specific, consider prey-predator equations of the form (of which equation (18) is an example)

$$dN_1/dt = N_1 f(N_1, N_2), \quad (20a)$$

$$dN_2/dt = N_2\, g(N_1). \quad (20b)$$

These reduce to a single equation for the prey population,

$$dN_1/dt = N_1 f\left(N_1(t), C \exp\left[\int^t g(N_1(t'))dt'\right]\right). \qquad (21)$$

In short, we will have an equation of the general character of equation (3) or equation (4), with a time delay roughly corresponding to the time for the predator population to respond to prey changes (if our single species equation describes the prey population), or for the vegetation to recover (if our equation describes the herbivores). This point is discussed more fully by Caughley (1975); we see that, in a very broad sense, the ideas developed in section 2 also include prey-predator situations.

Models intermediate between a single population (with all other environmental variables purely passive) and a fully 2-dimensional system have been studied by McMurtrie (1975) and others. Here one uses an equation such as the logistic equation (2), but acknowledges that the resources, K, are depleted to a degree which depends on the magnitude of the population, and that there is some characteristic time, τ, for the resource to recover. An example of such a K-value which depends on some average over past values of $N(t)$ is

$$K(t) = K_0 - b\int_{-\infty}^{t} N(t') \exp\left[-(t - t')/\tau\right]dt'. \qquad (22)$$

Here K_0 is the resource level in the absence of consumers, b is the rate of depletion of the resource by consumers, and τ is the resource regeneration time. A linearized stability analysis of equation (22) in conjunction with the logistic equation (2) shows that (like the ordinary logistic) the equilibrium point at $N^* = K_0/(1 + b\tau)$ is always stable, but that (unlike the ordinary logistic) perturbations die away in an oscillatory way once

$$b > (1 - r\tau)^2/(4r\tau^2). \qquad (23)$$

High depletion rates have a tendency to produce oscillations.

(ii) Discrete growth. As discussed for single populations, there will be situations where generations do not overlap, and the prey-predator dynamics are described by difference equations. Arthropod prey-predator systems provide many examples.

An interesting subclass of such models arises when each adult host (prey) in generation t either is parasitized to produce an adult parasite (predator) in generation $t + 1$, or escapes parasitism to produce F adult progeny. Denoting the host and parasite populations by H and P respectively, and assuming them to be temporally synchronized (see Oster, this volume), we have

$$P_{t+1} + H_{t+1}/F = H_t. \tag{24}$$

All details of the parasites' searching behavior are now contained in a function $f(H_t, P_t)$, which gives the fraction of hosts escaping parasitism,

$$H_{t+1} = F\,H_t\,f(H_t, P_t). \tag{25}$$

Equations (24) and (25) complete the description of the system. These host-parasite situations have received much attention, both in the field and in the laboratory (see, e.g., Varley, Gradwell and Hassell, 1974), and rightly so. They are much more amenable to study than most prey-predator systems, because all the biological details are in the functional response, $f(H, P)$, with the usual complications of the numerical response being avoided by the simple but realistic relation (24).

Initial theoretical studies (Nicholson and Bailey, 1935) used $f = \exp(-aP)$, giving a system which is unstable, with growing oscillations. Recent work has incorporated a variety of more realistic refinements (Hassell and May, 1973, 1974), possibly the most significant of which is spatial heterogeneity (which is outside the scope of this article). The effect of mutual interference between parasites, which increasingly diminishes their searching efficiency as their population density increases, is also important, and may be included by writing

$$f(H, P) = \exp(-QP^{1-m}). \tag{26}$$

Here m is a measure of the degree of mutual interference. On the one hand, the methods outlined in appendix I may now be used to show that the system defined by equations (24), (25) and (26) has a stable point iff

$$1 > m > 1 - (F - 1)/(F \ln F), \tag{27}$$

and has unboundedly growing oscillations otherwise. On the other hand, for many parasite populations the m-value of equation (26) may be deduced from laboratory or field data. This comparison between theoretical and empirical results suggests that parasite mutual interference is an important stabilizing influence in many real host-parasite systems.

This line of study has been pushed further by Rogers and Hubbard (1974), who argue that parasites with but few host species may be expected typically to have larger m-values than parasites with many alternate hosts: the latter may be able to stabilize their populations by switching among the different host species, while the former need to rely more on internal stabilizing mechanisms, such as mutual interference. The data summarized by Rogers and Hubbard (1974) (see Table 3) supports these notions. This is an interesting example of the way detailed studies of population dynamics may be bent toward such evolutionary questions as the evolution of monophagy.

In our discussion of simple deterministic models for single populations, we saw that continuous growth (differential equations) led to stable points or to stable limit cycles, while discrete growth (difference equations) led to a richer spectrum of behavior, with stable points, stable cycles, or chaos. The same is true for simple deterministic prey-predator models.

Although we have seen that prey-predator systems with continuous growth usually possess either a stable point or a stable limit cycle, a similarly generic discussion of discrete growth models is difficult. However, there is an illuminating investigation by Beddington, Free and Lawton (1975) of the particular system

$$H_{t+1} = H_t \exp\left[r(1 - H_t/K) - aP_t\right], \tag{28a}$$

$$P_{t+1} = \alpha H_t\left[1 - \exp(-aP_t)\right]. \tag{28b}$$

TABLE 3

Correlation between parasite mutual interference constant, *m*, and number of host species attacked by the parasite. (From Rogers and Hubbard, 1974.)

Parasite Species	*m*	Number of hosts attacked	Field or laboratory
Nasonia vitripennis (Walk.)	0.02	38	lab
Pteromalus puparum (L.)	0.00	20	lab
Nemeritis canescens (Grav.)	0.28	12	lab
Dahlbominus fuscipennis (Zett.)	0.66	8	lab
Cryptus inornatus (Pratt.)	0.38	5	lab
Chelonus texanus (Cress.)	0.53	5	lab
Cyzenis albicans (Fall.)	0.52	4	field
Telenomus nakagawai	0.48	3	field
Cratichneumon culex (Muell.)	0.87	3	field
Encarsia formosa	0.39	2	lab
Pseudeucoila bochei (Weld.)	0.69	2	lab
Apanteles fumiferanae	0.96	1	field

Numerical studies indeed show regimes of stable points, of stable limit cycles, and of chaos. The onset of chaotic behavior takes place at lower *r*-values than for the corresponding single species model. Beddington, Free and Lawton's paper is rich in fascinating detail, which is illustrated by photographs of oscilloscope trajectories. It is almost certain that other difference equation models for prey-predator systems will show a range of dynamical behavior similar to equation (28).

4. TWO POPULATIONS: COMPETITION

Models for two competitors were studied in the 1920's, and motivated Gause and later workers to undertake laboratory studies on the coexistence of species with "identical" resource requirements. More recently, theoretical work has tended to address the question of the limits to similarity between coexisting competitors (see, e.g., MacArthur, 1972), and it has been argued that environmental unpredictability (and therefore stochastic, rather than deterministic, models) may play an important part. These issues are succinctly covered in a short paper which I have reprinted here as appendix III; an account of the mathematical calculations which underpin this paper are in May (1974c).

It is to be emphasized that these estimates as to the limits to similarity among coexisting competitors are rough order-of-magnitude ones. The sign in the equations is most decidedly $\sim$, not $=$. Thus factor-of-2 differences between calculations based on differently shaped utilization functions, or on different criteria as to how small the population must be before it is highly likely to become extinct, are inconsequential. On the other hand, if the noise spectrum for the random environmental fluctuations has a high degree of correlation (either between different points in time, or between different points on the resource axis), a high degree of niche overlap may be possible, at least in principle. Arguments for and against the biological relevance of the models studied in appendix III are to be found in Schoener (1974), Feldman and Roughgarden (1975), Abrams (1975), Yorke (1975).

Deterministic differential equation models for two competitors can exhibit only stable points. But as soon as there are three or more competing populations, even the classical Lotka-Volterra differential equations can exhibit not only stable limit cycles, but also strange nonperiodic orbits ("strange attractors"): see May and Leonard (1975), Gilpin (1975). As ever, difference equation models for two competitors can manifest stable points, stable cycles, and chaos (May, 1974b; Hassell and Comins, 1975).

5. TWO POPULATIONS: MUTUALISM

Most contemporary ecology texts have chapters, complete with simple mathematical models, on competition and on prey-predator. With the outstanding exception of the second edition of Whittaker's (1975) book, analogous chapters on mutualism are lacking. I think the reasons for this are partly historical, and are partly because the simple, quadratically nonlinear, Lotka-Volterra models that capture some of the essential dynamical features of competition and prey-predator are inadequate for even a first discussion of mutualism, as they tend to lead to silly solutions in which both populations undergo unbounded exponential growth, in an orgy of mutual benefaction.

Minimally realistic models for two mutualists must allow for saturation in the magnitude of the reciprocal benefits, which tends to produce a stable point: see, e.g., Whittaker (1975, pp. 37–42) or the more general formulation in Hirsch and Smale (1974, ch. 12, problem 2).

Slightly more detailed models confirm the propensity for spatially homogeneous mutualistic associations to exhibit instability, particularly at low densities. To be specific, consider a model for the density of a plant population, N_1, and its obligate pollinator, N_2. If the nectar provided by the plant is the resource sustaining the pollinators, we may write a logistic equation

$$dN_2/dt = r\,N_2(1 - aN_2/N_1). \tag{29}$$

The plant population has some intrinsic death rate, $-dN_1$, and a "birth rate" which depends on the effective value of the pollinator population when this population is low, but saturates to some limit, I, set by environmental resource availability when pollinators are plentiful (cf. Schoener, 1973)

$$dN_1/dt = -dN_1 + I\big[\hat{N}_2/(\hat{N}_2 + C)\big]. \tag{30}$$

The effective pollinator density, $\hat{N}_2$, allows for the fact that at sufficiently low plant population densities relatively few pollinators will actually find plants (the number succeeding being proportional to some binary collision term, $\hat{N}_2 \sim N_2N_1/D$), whereas at high plant densities essentially all pollinators find plants: that is,

$$\hat{N}_2 = N_2\big[N_1/(N_1 + D)\big]. \tag{31}$$

Amalgamating equations (30) and (31), we have

$$\frac{dN_1}{dt} = N_1\left\{-d + \frac{I\,N_2}{N_1N_2 + N_1C + CD}\right\}. \tag{32}$$

Provided that $I > adC$, and that $(I - adC)^2 > 4ad^2CD$, the dynamical landscape of this mutualistic system is as illustrated in Fig. 6. These restrictions on the parameters very roughly require

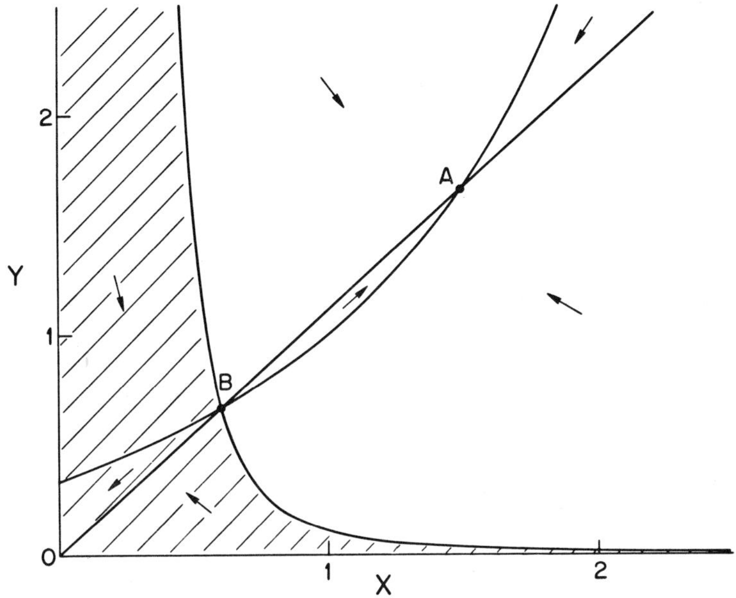

FIG. 6. Illustrating the dynamical properties of the pair of equations (29) and (32). Here $X = N_1/D$ is the dimensionless plant population, $Y = N_2/C$ the pollinator population, and the relevant combinations of parameters are $aC/D = 0.9$, $I/rD = 1.0$, $d/r = 1/3$. The features of this figure are as discussed in the text: trajectories originating outside the hatched region are attracted to the stable point A, while trajectories originating inside the hatched area are attracted to the origin.

that the plant "birth rate" has the capacity to outrun the death rate, and that D (the density at which plants become hard to find) is not too large: if the conditions are not met, both populations collapse to zero. In Fig. 6, the straight line through the origin is the isocline along which $dN_2/dt = 0$, and the curve intercepting it at the points A and B is the isocline $dN_1/dt = 0$. The directions in which population trajectories move in the various sectors are as shown by the arrows. The point A is a stable point, and attracts all trajectories originating outside the hatched region abutting the axes. The point B is unstable. All trajectories originating inside the hatched region are attracted to the origin, i.e., the populations become extinct.

If the system modeled by equations (29) and (32) is embedded in an environment which is subject to large amplitude perturbations, it is sooner or later liable to be swept into the hatched area in Fig. 6, whereupon extinction will probably follow. The long term persistence of mutualistic systems of this kind will be favored by environments which are stable and predictable. These dynamical considerations may go some way toward explaining why obligate mutualisms of this sort are less prominent in temperate and boreal ecosystems than in tropical ones. A good survey of empirical evidence bearing on this point is given in Farnworth and Golley (1974, pp. 29–31), where it is observed that not only are there no obligate ant-plant mutualisms north of 24°, no nectarivorous or frugivorous bats north of 33°, no orchid bees north of 24° in America, but also within the tropics mutualistic interactions are more prevalent in the warm, wet evergreen forests than in the cooler and more seasonal habitats.

The equation (32) is still overly simple, in that it contains no time lags. Although the disappearance of their plant resource will have an immediate effect on the pollinator population (equation (29)), there will usually be a substantial time lag between the disappearance of pollinators and the subsequent decline in the plant population (equation (32)). Inclusion of such time lags tends to make the system less responsive to perturbations, and is thus inclined to stabilize it. This abstract theoretical remark marches with an observation made by Wallace (1878), to the effect that one reason why herb species (e.g., members of the family Compositae) on isolated islands tend to evolve into shrubs or trees is that individuals thereby live longer and the population can survive a period without pollinators, diminishing the inherent instability of the system.

6. MANY INTERACTING POPULATIONS

We could now go on to study 3-dimensional systems of differential or difference equations, for three populations interacting in a variety of ways. Such models tend to be complicated, and do not commonly give clear biological insights.

Another approach is to study rather general multispecies models. These may be classes of specific models (e.g., multispecies Lotka-Volterra prey-predator models), or they may describe the community dynamics by a matrix of interaction coefficients in the neighborhood of equilibrium, or they may describe only the topological structure of the food web (i.e., only the signs of the interactions between the various species). At any level of abstraction, the dynamical properties of a model with many species and many interactions may be contrasted with the corresponding model with few species and few interactions. In this way we may hope to gain some understanding of the general mathematical relation between stability and complexity in model ecosystems. Such studies have been reviewed extensively elsewhere (May, 1975a), and I shall not recapitulate here.

Other aspects of the organization of multispecies communities may be approached from many different angles. For example, interesting problems arise in explaining the observed patterns of relative abundance of the various species in a community. Also of interest is the relation between the area of an island (actual or virtual) and the number of species on it: one may seek to derive such relations from patterns of species relative abundance, or from dynamical aspects of the number of species that can be "packed together" and persist, or simply on the basis of empirical regression relations that fit the data. Such species-area studies have been one of the more successful ventures in theoretical ecology, with applications ranging from environmental management to paleontology (as briefly reviewed by May (1975), and by Gould (1974), respectively).

The overall purpose of this article has been to argue the case that the methods of classical applied mathematics can make contributions toward understanding the world of plants and animals.

APPENDIX I: STABILITY CRITERIA

A linearized stability analysis of equation (3) may be carried out by writing $N(t) = K + x(t)$, expanding in power series in x, and discarding terms of order x^2 and higher:

$$dx/dt = -r\,x(t - T). \tag{A1}$$

Writing $x(t) = x(0)\exp(pt)$, we get

$$p = -r\,e^{-pT}. \tag{A2}$$

The asymptotic stability behavior will clearly be determined by the solution (eigenvalue), p_0, with the largest real part: if $\mathrm{Re}(p_0) > 0$, the point will be unstable; if $\mathrm{Re}(p_0) < 0$, it will be stable, with exponential damping if $\mathrm{Im}(p_0) = 0$ and oscillatory damping otherwise.

Thus the transition from stable to unstable point occurs when the dominant solution of equation (A2) is purely imaginary: $p_0 = iy$. Writing the real and imaginary parts of equation (A2) in explicit fashion, we have $0 = -r\cos(yT)$ and $y = r\sin(yT)$. The first of these gives $yT = \frac{1}{2}\pi$, and the second then gives the final result $rT = \frac{1}{2}\pi$ for the transitional parameter value. As p_0 crosses the imaginary axis (i.e., as rT increases beyond $\frac{1}{2}\pi$), the Hopf theorem tells us that the stable point will give way to a stable limit cycle.

Within the stable point regime, the transition from exponential to oscillatory damping occurs when it is no longer possible to find a purely real solution for p_0. By writing $p_0 = -x/T$, we see from equation (A2) that there will be exponential damping so long as it is possible to find a positive real solution, x, to the equation $xe^{-x} = rT$: since the left hand side of this equation has a maximum value of e^{-1}, we see that there will be exponential damping iff $rT < e^{-1}$.

The corresponding linearized stability analysis of the more general equation (4) has been set out elsewhere (May, 1973). It is not essentially different from that above. One first linearizes about equilibrium to get

$$dx/dt = -r\,N^* \int_{-\infty}^{t} Q(t - t')x(t')dt', \tag{A3}$$

and then takes a Laplace transform to arrive at the relation

$$p + r\,\tilde{Q}(p)/\tilde{Q}(0) = 0. \tag{A4}$$

Here $\tilde{Q}(p)$ is the usual Laplace transform of $Q(t)$. The transition from exponentially damped to oscillatorily damped stable point takes place as the dominant root, p_0, splits off the negative real axis, and the Hopf bifurcation from stable point to stable limit cycle takes place as p_0 traverses the imaginary axis. In short, the behavior exemplified by equation (3) (where $Q(t) = \delta(t - T)/K$ and $\tilde{Q}(p) = \exp(-pT)/K$) is generic, and may be followed through in detail for various other choices for $Q(t)$.

The analysis may be extended, mutatis mutandis, to the general delay-difference equation (8). The usual linearization about equilibrium, $N_t = N^* + x_t$, leads to

$$x_{t+1} = x_t - b\,x_{t-T}. \tag{A5}$$

Here b is defined by equation (9). Time is now a discrete, not a continuous, variable and we may write $x_t = x_0(\lambda)^t$, to get

$$\lambda = 1 - b\lambda^{-T}. \tag{A6}$$

The stability is now set by the eigenvalue(s) with the largest modulus, λ_0: the point N^* will be locally stable iff $|\lambda_0| < 1$, and it will be exponentially, rather than oscillatorily, damped iff, in addition, λ_0 is real and positive. The sorts of arguments used earlier in this appendix may then be used to derive the criteria (10) and (11).

APPENDIX II: BIFURCATIONS AND DYNAMIC COMPLEXITY

This appendix outlines the essential process underlying the successive bifurcations illustrated in Fig. 3. A much more detailed exposition is in May and Oster (1975).

Consider the general difference equation

$$N_{t+1} = F(N_t). \tag{A7}$$

Populations which are k time steps apart are related by the k-times-composed map $F^{(k)}$:

$$N_{t+k} = F\big(F(\cdots F(N_t)\cdots)\big) \equiv F^{(k)}(N_t). \qquad \text{(A8)}$$

Clearly, fixed points of period k, $N_i^{*(k)}$ ($i = 1, 2, \ldots, k$), are found where the map $F^{(k)}$ intersects the 45° line, $N_{t+k} = N_t$:

$$N^{*(k)} = F^{(k)}(N^{*(k)}). \qquad \text{(A9)}$$

The local stability of these period k points depends on the slope of $F^{(k)}$, defined as

$$\lambda^{(k)} = \big[dF^{(k)}(N)/dN\big]_{N=N_i^{*(k)}}. \qquad \text{(A10)}$$

That is, using the expansion (A8) for $F^{(k)}$,

$$\lambda^{(k)} = \prod_{i=1}^{k} \big[dF(N)/dN\big]_{N=N_i^{*(k)}}. \qquad \text{(A11)}$$

Note that the slope $\lambda^{(k)}$ is the same for all period k points. By the methods outlined in appendix I, these period k fixed points are stable iff $1 > \lambda > -1$.

The central question is, starting with a stable cycle of period k, what happens as the local map $F^{(k)}$ steepens so that $\lambda^{(k)}$ steepens beyond -1, producing instability? This situation is illustrated on the left hand side in Fig. 7.

To answer this question, we turn to the $2k$-times-composed map, $F^{(2k)}$. The period k points comprise a degenerate set of period $2k$ points, and the slope of the map $F^{(2k)}$ at these points is

$$\lambda^{(2k)} = \big[dF^{(2k)}(N)/dN\big]_{N=N_i^{*(k)}}. \qquad \text{(A12)}$$

The expansion which led to equation (A11) now produces

$$\lambda^{(2k)} = \left\{ \prod_{i=1}^{k} \big[dF(N)/dN\big]_{N=N_i^{*(k)}} \right\}^2. \qquad \text{(A13)}$$

That is

$$\lambda^{(2k)} = [\lambda^{(k)}]^2. \qquad \text{(A14)}$$

The fact that $\lambda^{(2k)}$ is the square on $\lambda^{(k)}$is the key to what follows. As the period k solution remains stable, with $\lambda^{(k)}$ steepening from 0 toward -1, the map $F^{(2k)}$ has slope $\lambda^{(2k)}$ less than (but approaching) $+1$, and it only cuts the 45° line once in the neighborhood of $N^{*(k)}$. But as $\lambda^{(k)}$ steepens beyond -1, the map $F^{(2k)}$ has slope $\lambda^{(2k)}$ greater than $+1$, and it crosses the 45° line to cut it three times in the neighborhood of $N^{*(k)}$. Exactly at the juncture where the period k solution becomes unstable there is born a new (and locally stable) solution of period $2k$. Fig. 7 aims to illustrate this.

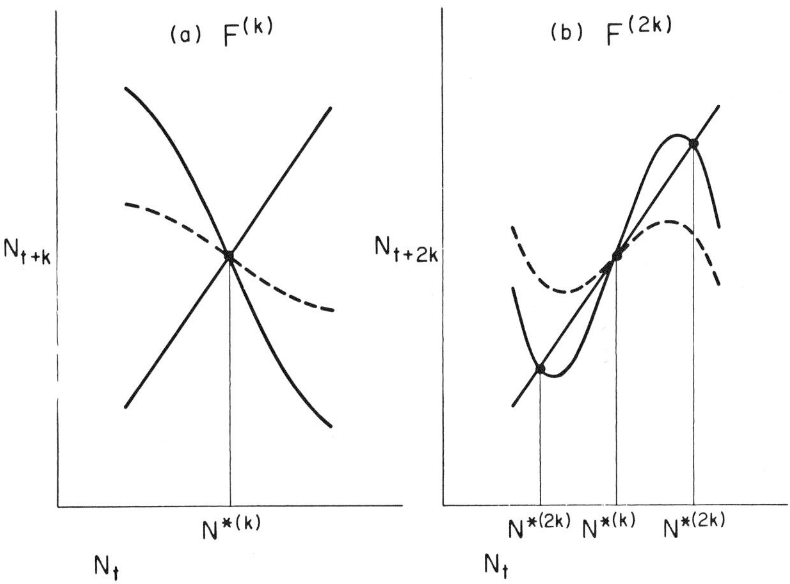

FIG. 7.(a) The dashed curve illustrates the map of $F^{(k)}$ (relating N_{t+k} to N_t) in the neighborhood of one of its period k fixed points, $N^{*(k)}$; here the slope of $F^{(k)}$ at its intersection with the 45° line, $\lambda^{(k)}$, has $|\lambda^{(k)}| < 1$, and the point is stable. For the solid curve $|\lambda^{(k)}| > 1$, and the fixed point becomes unstable. (b) This figure illustrates the corresponding maps of $F^{(2k)}$ (relating N_{t+2k} to N_t), in the neighborhood of the same period k fixed point, $N^{*(k)}$. The consequent features are as discussed in the text.

In particular, this generic argument may be applied to the general equation (A1), or to the specific equations (6) and (7), to see how the stable fixed point of period 1 gives rise, by successive bifurcations, to fixed points of period 2, 4, 8, 16, . . . , 2^n. A systematic approximation procedure for calculating the limiting parameter value, r_C, is given by May and Oster (1975).

APPENDIX III: NICHE OVERLAP AS A FUNCTION OF ENVIRONMENTAL VARIABILITY*

One of the central concepts in ecology is the competitive exclusion principle, which forbids the coexistence of two or more species making their livings in identical ways. Recently, an increasing amount of attention has been paid to the questions: How similar can competing species be if they are to remain in an equilibrium community? How identical is "identical"? How close can species be packed in a natural environment?

An answer to these questions may begin by noticing that in laboratory experiments, where the environment can be carefully kept unvarying, species whose ecology is well-nigh identical have coexisted for long periods [1]. A conjecture [2, 3] is that in the real world, environmental fluctuations will put a limit to the closeness of species packing compatible with an enduring community, and that species will be packed closer or wider as the environmental variations are smaller or larger.

Motivated by these ideas, we consider a one-dimensional resource spectrum, sustaining a series of species, each of which has a preferred position in the spectrum, and a characteristic variance about this mean position, as given by some "utilization function" (see Fig. 1). For example, the resource spectrum may be food size, and the consumers may be bird species each having a utilization function that describes their mean food size and its variance. The dynamics of this situation may be plausibly modeled by a system of first-order differential equations, with competition coefficients

*This article, by Robert M. May and Robert H. MacArthur, is reprinted from *Proc. Nat. Acad. Sci., USA*, **69** (1972), 1109–1113.

that depend on how closely species are packed; that is, on the degree of niche overlap (on the ratio of d to w in Fig. 1).

In the stability analysis of such models, two qualitatively different circumstances need be distinguished. In the unrealistic case when all the environmental parameters are strictly constant (deterministic), then in general the system remains stable even if an arbitrarily large number of species are packed in, arbitrarily close. On the other hand, when the relevant environmental parameters fluctuate (stochastic environment), there is a limit to the niche overlap consistent with long-term stability.

However, this limit to species packing depends on the environmental variance in a far-from-obvious and extremely interesting way (Fig. 3). If the fluctuations in the resource spectrum are severe, having variances comparable in magnitude with their mean values, the species packing is indeed roughly proportional to the environmental variance, as one would expect intuitively. But for fluctuations ranging from moderate to exceedingly small, the species packing attains an effective limiting value roughly equal to the width of the utilization functions. Thus, as the ratio between the variance and mean value in the resource spectrum, or other pertinent environmental parameter, falls from 0.3 to 0.0001, the

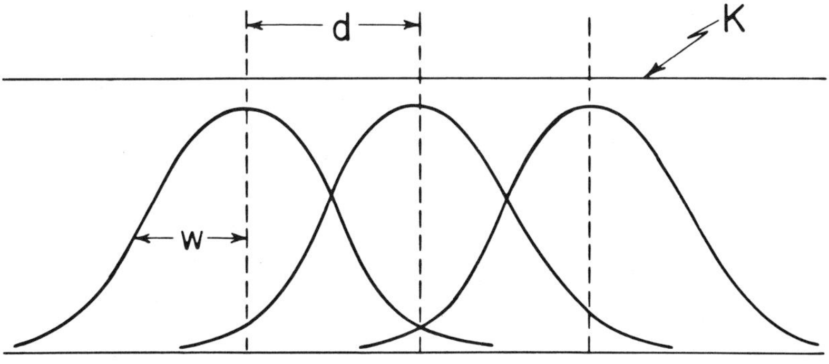

FIG. 1. The curve labeled K represents some resource continuum—say amount of food as a function of food size—that sustains various species whose utilization functions (characterized by a standard deviation w and a separation d) are as shown.

closest species packing consistent with stability falls only from 2 to 1 times the utilization function variance. Moreover, our general result is a robust one, being rather insensitive to the details of the mathematical model.

Collecting these statements, we observe that the species packing parameter d indeed goes to zero when the environmental variance becomes strictly zero, but that for any finite environmental variance, d remains roughly equal to the utilization function width, w. This result, which at first glance seems odd, reflects the technical fact that the mathematics contains an essential singularity around $d = 0$ (Eq. (6) and Fig. 2), so that there is a qualitative difference between an environmental variance that is small but finite, and one that is zero.

Following Hutchinson's [4] initial observations, Mac Arthur [3] has recently reviewed a body of semiquantitative work bearing on

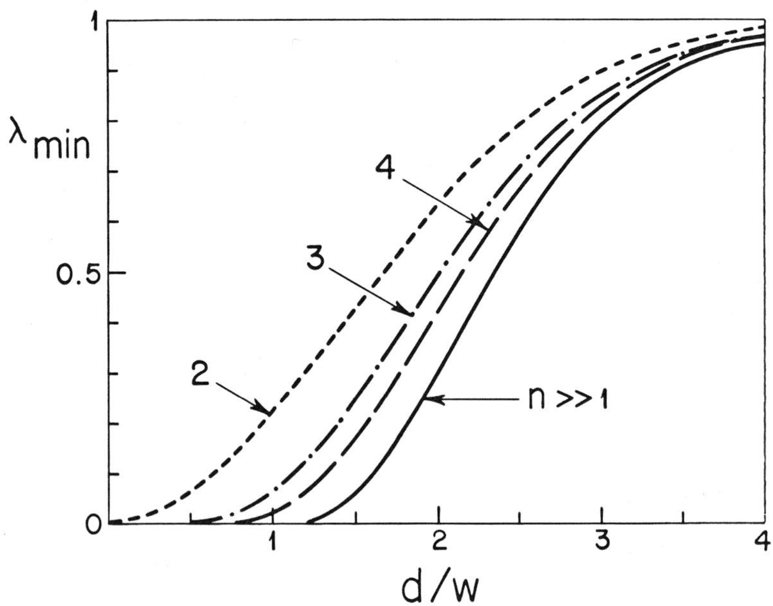

FIG. 2. The minimum eigenvalue of the stability matrix (Eq. (4)) as a function of niche overlap, d/w, for an n-species guild, where $n = 2, 3, 4$ and $n \gg 1$.

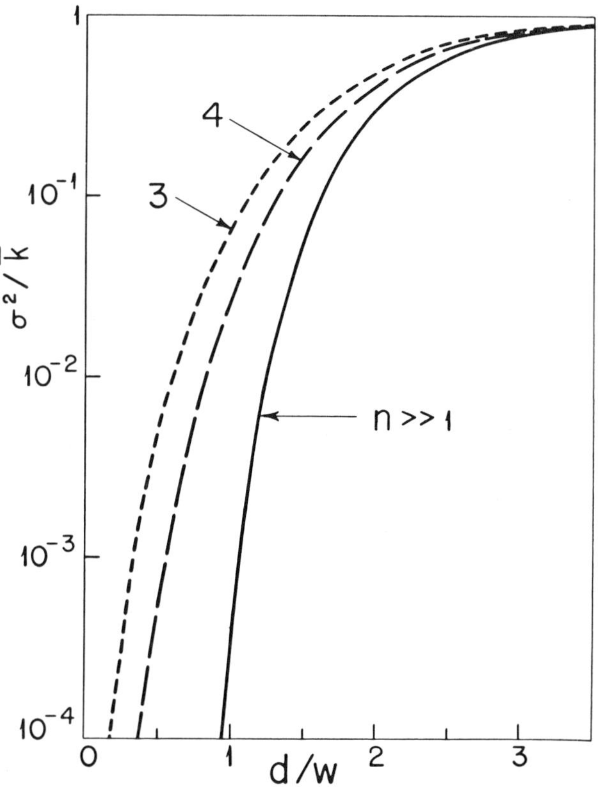

FIG. 3. The closest niche overlap, d/w, consistent with community stability in a randomly varying environment, whose fluctuations are characterized by a variance (relative to the mean) of $\sigma^2/\bar{k}$. The variance is plotted on a logarithmic scale to emphasize that, over a wide range, it has little influence on the species packing distance for $n > 2$.

species packing and character displacement among competing species. These empirical data, which are discussed more fully in Section IV, match the conclusions drawn from our model.

Two corollaries are worth mentioning.

First, most of the ideas advanced to account for the gradient in species diversity as one goes from the tropics to the poles may be summarized under three headings [1, 5]: (i) as a matter of history, there has been more time for speciation in the tropics than most other places; (ii) *total* niche "volume" is greater in the tropics, which tend to be more productive, less seasonal, and more floristically complex, both in stratification and diversity; (iii) more niche overlap is permitted in the tropics by the unvarying environment. The potential number of species is the total volume [i.e., (ii)] divided by the effective niche volume per species [i.e., (iii)], and this potential will be realized if enough time is available [i.e., (i)]. The intuitive basis for the argument (iii) was set out in the second paragraph of this introduction, but the quantitative conclusion that niche overlap is only weakly dependent on the degree of environmental fluctuation (unless very severe) suggests that (iii) is a relatively unimportant factor in explaining the species diversity gradient, at least until one gets to extreme latitudes. It should be emphasized that our conclusion that species packing, d, is roughly proportional to utilization function width, w, implies only that *niche overlap* is largely independent of the environmental variance, σ^2. It remains true that the total number of species packed into an interval on the resource spectrum is greater if they are specialists (small w) than if they are generalists (large w); the question as to what ultimately determines w remains open.

Second, in this model, which explicitly treats only one trophic level, it is obvious that greater complexity (in the form of more species, more closely packed) makes for lesser stability. In a perfectly stable deterministic environment, arbitrarily close packing and rich speciation is possible, and to a certain limited extent the greater the environmental steadiness, the closer the packing, and the richer the consequent assembly of species. Insofar as this example adds a piece to the complexity-stability jigsaw puzzle, it is that complexity is a fragile thing, permitted in this instance by environmental steadiness: this is quite the opposite of the conventional "complexity begets stability" wisdom [6].

The details of the model are outlined in Section I, and the results derived in Section II. Section III contains a short account of work bearing on the insensitivity of the main results to the details of the model.

I. THE MODEL DEFINED

Suppose one has a one-dimensional continuum of resources, such as food size, or vertical habitat, or horizontal habitat, that may be schematically depicted as in Fig. 1, where the curve labeled K shows amount of food as a function of food size, or amount of habitat as a function of height, and, in general, amount of resource as a function of x. Suppose further that this resource sustains various species, each of which has a utilization function $f(x)$ as depicted in Fig. 1, which characterizes the species' use of the resource spectrum. In particular, we note the mean position and the standard deviation, w, about this mean for the various species; i.e., the mean and the variance of the food size, or of the habitat height, etc. The separation, d, between the mean positions of species that are adjacent on the resource continuum will clearly be a measure of how densely the species are packed.

Mac Arthur [3, 7] has established a criterion that ensures that the actual community utilization of the resource will provide the best least-squares fit to the available resource spectrum. This requires the populations of the n species, $N_i(t)$ [labeled sequentially $i = 1, 2, \ldots, n$], to obey

$$\frac{dN_i(t)}{dt} = N_i(t)\left[k_i - \sum_{j=1}^{n} \alpha_{ij} N_j(t)\right] \tag{1}$$

where the k_i are integrals with respect to x over the product of the resource spectrum and the utilization function of the ith species, and the competition coefficients α_{ij} are convolution integrals between the utilization functions of the ith and jth species. With this, we are assured both that the equilibrium populations (obtained by setting all $d/dt = 0$) minimize the squared difference between available and actual "production," and also that nonequilibrium initial populations will move in time towards this minimum configuration.

Eq. (1) is, of course, the Lotka-Volterra competition equation, but tied to the underlying model illustrated by Fig. 1, so that we have explicit recipes for the k_i and α_{ij} in terms of direct biological assumptions. Specifically, if we assume that all the species' utilization functions are the usual bell-shaped gaussian curves, with common width w, and that they are uniformly spaced along the resource continuum (common d), the competition coefficients are

$$\alpha_{ij} = (w^2\pi)^{-\frac{1}{2}} \int_{-\infty}^{\infty} dx \exp\left[-\frac{x^2}{2w^2} - \frac{(x-(i-j)d)^2}{2w^2} \right]$$

$$= [\alpha]^{(i-j)^2}, \tag{2}$$

where we have for notational convenience defined

$$\alpha = \exp(-d^2/4w^2). \tag{3}$$

Quite apart from the teleology implicit in the assumption that communities minimize anything, a choice of fit other than least-squares will lead to equations superficially different from (1): however, their competition matrix characterizing small displacements from equilibrium will end up similar to that given below. As Lotka [8], and others since, have emphasized, Eq. (1) represents the first term in a Taylor expansion of a much wider class of equations, and thus should be useful in discussing the stability of potential equilibria.

In the stability analysis of equations such as (1), we first find the equilibrium populations, N_i^*, and then study small-amplitude perturbations by linearizing about this equilibrium. As a further simplification in our model, we rather arbitrarily choose the resource spectrum to be such that the community best-fit to it (i.e., the equilibrium community) has *equal* populations for all species; for a large number of species, $n \gg 1$, this means a flat resource spectrum. The conventional analysis then shows the stability of the system to be given simply by the eigenvalues of the $n \times n$ competition matrix A, which from (2) has the form

$$A = \begin{bmatrix} 1 & \alpha & \alpha^4 & \alpha^9 & \cdot & \cdot & \alpha^{n^2} \\ \alpha & 1 & \alpha & \alpha^4 & \cdot & \cdot & \cdot \\ \alpha^4 & \alpha & 1 & \alpha & \cdot & \cdot & \cdot \\ \alpha^9 & \alpha^4 & \alpha & 1 & \cdot & \cdot & \cdot \\ \cdot & \cdot & \cdot & \cdot & \cdot & \cdot & \cdot \\ \alpha^{n^2} & \cdot & \cdot & \cdot & \cdot & \cdot & 1 \end{bmatrix}. \tag{4}$$

In short, in this section we have made several particular assumptions, which have given a specific form for the competition matrix, namely (4). Indeed, this is a form that can be plausibly justified from quite *ad hoc* considerations. In general, the system stability, and hence the permissible degree of niche overlap, hinges upon some such competition matrix. Other assumptions could give other (but similar) matrices, and the extent to which our conclusions are or are not tied up with the specific model outlined here is discussed in Section III.

II. THE MODEL ANALYZED

Deterministic environment. As just described, all the parameters in our model system are unvarying constants. Consequently the equilibrium configuration is stable, with perturbations damping out, so long as all the eigenvalues, λ, of $\mathbf{A}$ are positive (notice that a minus sign was absorbed in the definition of the competition matrix). But $\mathbf{A}$ is a positive definite form for all $0 \leqslant \alpha < 1$ (i.e., for all d, see Eq. (3)), with the consequence that stability sets no limit to the species packing in a strictly deterministic environment. Moreover, in general the more species packed in, the better the least-squares fit to the resource spectrum.

Nevertheless it is interesting to see how the *smallest* eigenvalue of $\mathbf{A}$, which sets the stability character, varies with niche overlap, as measured by d/w. For $n \gg 1$, we have (see Appendix)

$$\lambda_{\min} = 1 - 2\alpha + 2\alpha^4 - 2\alpha^9 + 2\alpha^{16} - \cdots . \tag{5}$$

This series may be summed, by an elegant method, to get an approximation that is very accurate unless $d \gg w$ (see Appendix):

$$\lambda_{\min} = 4\pi^{\frac{1}{2}}(w/d)\exp\left[-\pi^2 w^2/d^2\right]. \tag{6}$$

This is a remarkable result. For substantial niche overlap, i.e., d/w small, $\lambda_{\min}$ tends to zero faster than any finite power of d/w: there is an essential singularity at $d/w = 0$. Thus, although $\lambda_{\min}$ is indeed necessarily positive even for small d/w, it becomes exceedingly tiny, corresponding to extremely long damping times. This result foreshadows the results below.

Fig. 2 illustrates Eq. (6), along with numerical results for $n = 2$, 3, 4. Notice that for practical purposes, $n = 4$ is hard to distinguish from "$n = \infty$".

Stochastic environment. More realistically, there will be random environmental fluctuations, so that the resource continuum will be noisy. This means the quantities k_i in Eq. (1) will not be constants, but rather will be random variables. We assume

$$k_i = \bar{k}_i + \gamma_i(t) \tag{7}$$

where $\bar{k}_i$ is the constant mean value (having the common value $\bar{k}$ for large n), and $\gamma_i(t)$ is gaussian "white noise", with variance measured by σ^2.

In this stochastic problem, we may no longer talk of *the* species populations, but only of their joint probability distribution. To a good approximation, this is a multivariate normal distribution in the fluctuations about the means, and the probability of any species becoming extinct will be small (corresponding to the mechanical "stability" of the deterministic case) if the smallest eigenvalue of the competition matrix **A** roughly obeys

$$\lambda_{\min} > \sigma^2/\bar{k}. \tag{8}$$

This result [9] is commonsensical. In a randomly fluctuating environment, it is not enough that all the eigenvalues be positive, but rather they should be bounded away from zero by an amount roughly proportional to the environmental noise level.

Combining the qualitative equation, (8), with Fig. 2, we arrive at an estimate of the closest species packing, d/w, consistent with stability for a given environmental noise level, $\sigma^2/\bar{k}$. These results, illustrated in Fig. 3, are as discussed in the Introduction.

In particular, we see explicitly from Eq. (6) that for large n this closest degree of niche overlap depends on the environmental fluctuations only as $\sqrt{\ln \sigma^2}$, a very weak dependence. The results for $n = 3, 4$, although allowing a slightly closer limiting packing distance, display a similar insensitivity to the degree of random fluctuation, so long as it is not severe.

III. HOW ROBUST ARE THESE RESULTS?

The question arises, to what extent are these results peculiar to our particular model? We catalogue some answers.

(i) We chose gaussian utilization functions. Alternative $f(x)$ ranging from the opposite extremes of rectangles through to back-to-back exponentials or Lorentz lineshapes lead to **A** matrices different from (4), but the plot of $\lambda_{\min}$ as a function of d/w retains the essential features of Fig. 2 in all cases.

(ii) We chose the width and separation of the utilization functions to be constant. If the width w changes in some systematic way along the resource continuum, our results are preserved, as long as the separation d changes in the same proportion, keeping d/w roughly constant.

(iii) The resource spectrum of Fig. 1 was assumed to be such that, at equilibrium, all populations are equal. Extensive investigation of various resource spectrum shapes for $n = 2$, 3, and 4 suggests that our results are not dependent on this feature, so long as all species are present in significant numbers in the equilibrium community.

(iv) The implications of use of Eq. (1) were discussed in section I.

(v) The stochasticity of the environment was taken to be gaussian "white noise," i.e., no correlation between the fluctuations at successive instants. In practice, this means only that fluctuations be correlated over times short compared to all other relevant time scales in the system [9].

(vi) Our model is for competition in one resource dimension. Cody's [10] classification of partitioning in the three-resource dimensions of horizontal habitat, vertical habitat, and food for 10 grassland bird communities around the world shows eight of them to be organized largely in one dimension (food selection), so that our model is not wholly unreasonable. Moreover, the model is directly relevant to niche overlap in two or more orthogonal resource dimensions, and may even be useful as a metaphor for more complicated circumstances.

IV. COMPARISON WITH REAL ECOSYSTEMS

In a classic paper, Hutchinson [4] observed that in various circumstances, including both vertebrate and invertebrate forms, character displacement among sympatric species leads to sequences in which each species is roughly twice as massive as the next; i.e., linear dimensions as measured by bills or mandibles in the ratio 1 · 2–1 · 4. MacArthur's more recent and quantitative reviews [3] of such data point to there being a limiting value to niche overlap in the natural world, corresponding to d/w in the range 1–2. Also pertinent is Simpson's [11] review of the factors making for latitudinal and altitudinal species diversity gradients among North American mammals: it concludes that degree of niche overlap is not an important contributing factor.

The work that seems to come closest to our one-dimensional model is that of Terborgh, Diamond, and Beaver on various guilds of birds in an assortment of habitats that have various degrees of environmental stability. Even so, such comparisons with the theory are necessarily approximate, partly because our α (Eq. (3)) comes from utilization functions that are not just percentage of time or of diet, but rather have weighting terms for resource renewal [3, 7]: all available information from nature contains unweighted utilizations.

Terborgh [12] has shown five species of tropical antbird, segregating by foraging height, have mean heights separated by one standard deviation; i.e., $d/w \sim 1$. MacArthur's analysis [3] of Storer's data [13] on the food weight distribution for three congeneric species of hawks also leads to $d/w \sim 1$. Diamond's [14] extensive data on weights of tropical bird congeners that sort out largely (but not wholly, so that d/w should be smaller than our one-dimensional theory predicts) by size differences leads to weight ratios around $1 \cdot 6$–$2 \cdot 3$; when Hespenheide's analysis [15] of the relation between weight ratio and α is used, Diamond's results become $\alpha \cong 0 \cdot 8$–$0 \cdot 9$, i.e., $d/w \cong 0 \cdot 6$–$1 \cdot 0$. In the Sierra Nevada, Beaver (personal communication) has shown that species packing in a brushland bird community appears equal to that in a forest foliage gleaning guild, although the microenvironment is thought to be significantly more unvarying in the forest.

In brief, the basic conclusion that emerges in a nonobvious but robust way from our mathematical model, namely that there is a limit to niche overlap in the natural world and that this limit is not significantly dependent on the degree of environmental fluctuation (unless it be severe, as in the arctic), seems to be in harmony with such facts as are known about real ecosystems.

APPENDIX

For large n, where "end effects" at the extremes of the resource spectrum are unimportant, we may pretend that the resource continuum is cyclic (so that the species labeled 1 adjoins that labeled n), whereupon the competition matrix $\mathbf{A}$ of Eq. (4) is slightly modified to become related to a class of matrices discussed by Berlin and Kac [16]. Using their approach, one can obtain Eq. (5). That this trick of imposing artificial cyclic boundary conditions does not alter the eigenvalues for $n \gg 1$ is a point made clear in the literature on the physicists' Ising model, from which comes Berlin and Kac's paper.

The series in Eq. (6) is identically equal to the contour integral

$$\frac{1}{2i} \oint_C \frac{\exp\,(z^2 \ln \alpha)\,dz}{\sin\,(\pi z)} \,.$$

Here the contour C encloses all poles of the integrand up to $z = \pm\, n$, where the series has n terms. An $n \to \infty$, C is the circle at infinity in the complex plane. Using the standard Jordan contour, and ignoring correction terms of relative order exp $(-4\pi^2 w^2/d^2)$, which are thoroughly negligible for $d/w < 3$ or so, we arrive neatly at Eq. (6).

At the other extreme, for $n = 2$ the eigenvalues of **A** are easily found directly. For other finite values, such as $n = 3, 4$, we take a meat axe and display λ_{min} as a numerical function of d/w.

REFERENCES TO THE NICHE OVERLAP ARTICLE

1. R. S. Miller, *Advances in Ecological Research*, Vol. 4, Academic Press, New York, 1967, pp. 1–74 (see pp. 35–46).
2. ———, *Advances in Ecological Research*, Academic Press, New York, 1967, p. 67.
3. R. H. MacArthur, *Avian Biology*, Vol. 1, Academic Press, New York, 1971, pp. 189–221; *Geographical Ecology*, Harper and Row, New York, 1972, (in press).
4. G. E. Hutchinson, *Amer. Natur.*, **93** (1959), 145–159.
5. P. H. Klopfer, *Behavioral Aspects of Ecology*, Prentice-Hall, Englewood Cliffs, 1962, chap.3; E. R. Pianka, *Amer. Natur.*, **100** (1966), 33–46; R. H. MacArthur, *Biol. J. Linn. Soc.*, **1** (1969), 19–30; *Diversity and Stability in Ecological Systems*; *Brookhaven Symposium in Biology No.* 22, Nat. Bur. Standards, Springfield, Va., 1969.
6. R. M. May, *Math. Biosci.*, **12** (1971), 59–79.
7. R. H. MacArthur, *Proc. Nat. Acad. Sci. USA*, **64** (1969), 1369–1371; *Theor. Pop. Biol.*, **1** (1970), 1–11.
8. A. J. Lotka, *Elements of Physical Biology*, Williams and Wilkins, Baltimore, 1925, p. 62.
9. R. M. May, *Proc. Ecol. Soc. Aust.*, 1971, (in press); K. J. Astrom, *Introduction to Stochastic Control Theory*, Academic Press, New York, 1970; Z. M. Sykes, *J. Amer. Statist. Assoc.*, **64** (1969), 111–130.
10. M. Cody, *Amer. Natur.*, **102** (1968), 107–148.
11. G. G. Simpson, *Syst. Zool.*, **13** (1964), 57–73.
12. J. Terborgh, quoted in R. H. MacArthur, 1972: *Geographical Ecology*, Harper and Row, New York, 1972, Figure 6–4
13. R. W. Storer, *Auk*, **83** (1966), 423–436.
14. J. M. Diamond, *The Avifauna of the Eastern Highlands of New Guinea*, publ. Nuttall Ornithol. Club, Cambridge, Mass., 1972, (in press).
15. H. A. Hespenheide, *Ibis*, **113**, (1971), 59–72.
16. T. H. Berlin and M. Kac, *Phys. Rev.*, **86** (1952), 821–835.

REFERENCES TO MATHEMATICAL ASPECTS OF THE DYNAMICS OF ANIMAL POPULATIONS

1. P. Abrams, "Niche overlap and environmental variability," *Math. Biosci.*, **28** (1976), 357–372.
2. D. Auslander, G. F. Oster, and C. Huffaker, "Dynamics of interacting populations," *J. Franklin Inst.*, **297** (1974), 345–375.
3. W. Baltenzweiler, "The relevance of changes in the composition of larch bud moth populations for the dynamics of its numbers," *Dynamics of Populations* (eds: P. J. den Boer and G. R. Gradwell), Center for Agricultural Publishing, Wageningen, 1971, 208–219.
4. M. S. Bartlett, "Equations and models of population change," *The Mathematical Theory of the Dynamics of Biological Populations*, (eds: M. S. Bartlett and R. W. Hiorns), Academic Press, New York, 1973.
5. J. R. Beddington and R. M. May, "Time delays are not necessarily destabilizing," *Math. Biosci.*, **27** (1975), 107–119.
6. J. R. Beddington, C. A. Free, and J. H. Lawton, "Dynamic complexity in predator-prey models framed in difference equations," *Nature*, **255** (1975), 58–60.
7. G. Caughley, "Wildlife management and the dynamics of ungulate populations," *Advances in Appl. Biol.*, **1** (1976), 183–246.
8. H. Chitty, "Canadian arctic wildlife enquiry, 1945–49, with a summary of results since 1933," *J. Anim. Ecol.*, **19** (1950), 180–193.
9. H. N. Comins and D. W. E. Blatt, "Prey-predator models in spatially heterogeneous environments," *J. Theor. Biol.*, **47** (1974), 75–84.
10. R. Dajoz, *Dynamique des Populations*, Masson et cie, Paris, 1974.
11. C. Elton, *Voles, Mice and Lemmings: Problems in Population Dynamics*, Oxford University Press, Oxford, 1942.
12. E. G. Farnworth and F. B. Golley (eds.), *Fragile Ecosystems: Evaluation of Research and Applications in the Neotropics*, Springer-Verlag, New York, 1974.
13. M. W. Feldman and J. Roughgarden, "A population's stationary distribution and chance of extinction in a stochastic environment," *Theor. Pop. Biol.*, **7** (1975), 197–207.
14. M. E. Gilpin, "Limit cycles in competition communities," *Amer. Natur.*, **109** (1975), 51–60.
15. S. J. Gould, "The great dying," *Nat. Hist.*, *LXXXIII*: **8** (1974), 22–27.
16. M. P. Hassell and H. N. Comins, "Discrete time models for two-species competition," *Theor. Pop. Biol.*, **9** (1975), 202–221.
17. M. P. Hassell and R. M. May, "Stability in insect host-parasite models," *J. Anim. Ecol.*, **42** (1973), 693–726.

18. ———, "Aggregation of predators and insect parasites and its effect on stability," *J. Anim. Ecol.*, **43** (1974), 567–594.

19. M. P. Hassell, J. H. Lawton, and R. M. May, "Patterns of dynamical behavior in single-species populations," **45** (1976), 471–486.

20. M. W. Hirsch and S. Smale, *Differential Equations, Dynamical Systems, and Linear Algebra*, Academic Press, New York, 1974.

21. G. E. Hutchinson, "Circular causal systems in ecology," *Ann. N.Y. Acad. Sci.*, **50** (1948), 211–246.

22. A. R. Kiester and R. Barakat, "Exact solutions to certain stochastic differential equation models of population growth," *Theor. Pop. Biol.*, **6** (1974), 199–216.

23. A. N. Kolmogorov, "Sulla teoria di Volterra della lotta per l'esistenza," *Giorn. Istituto Ital. Attuari*, **7** (1936), 74–80.

24. C. J. Krebs, *Ecology: The Experimental Analysis of Distribution and Abundance*, Harper and Row, New York, 1972.

25. S. A. Levin and R. M. May, "A note on difference-delay equations," *Theor. Pop. Biol.*, **9** (1976), 178–187.

26. S. A. Levin and R. T. Paine, "Disturbance, patch formation and community structure," *Proc. Nat. Acad. Sci.*, **71** (1974), 2744–2747.

27. R. Levins, "The effect of random variations of different types on population growth," *Proc. Nat. Acad. Sci.*, **62** (1969), 1061–1065.

28. R. C. Lewontin and D. Cohen, "On population growth in a randomly varying environment," *Proc. Nat. Acad. Sci.*, **62** (1969), 1056–1060.

29. T-Y. Li and J. A. Yorke, "Period three implies chaos," *Amer. Math. Monthly*, **82** (1975a), 985–992.

30. ———, "Ergodic transformations from an interval into itself," preprint, 1975b.

31. D. Ludwig, *Stochastic Population Theories: Lecture Notes in Biomathematics*, vol. 3, Springer-Verlag, New York, 1974.

32. R. H. MacArthur, *Geographical Ecology*, Harper and Row, New York, 1972.

33. R. M. May, "Limit cycles in predator-prey communities," *Science*, **177** (1972), 900–902.

34. ———, "Time-delay versus stability in population models with two and three trophic levels," *Ecology*, **54** (1973), 315–325.

35. ———, "Ecosystem patterns in randomly fluctuating environments," *Progress in Theoretical Biology*, (eds: R. Rosen and F. Snell), Academic Press, New York, 1974a, 1–50.

36. ———, "Biological populations with nonoverlapping generations: stable points, stable cycles, and chaos," *Science*, **186** (1974b), 645–647.

37. ———, "On the theory of niche overlap," *Theor. Pop. Biol.*, **5** (1974c), 297–332.

38. ———, *Stability and Complexity in Model Ecosystems*, second ed., Princeton University Press, Princeton, 1975a.

39. ———, "Biological populations obeying difference equations: stable points, stable cycles, and chaos," *J. Theor. Biol.*, **51** (1975b), 511–524.

40. ———, "Island biogeography and the design of wildlife preserves," *Nature*, **254** (1975c), 177–178.

41. R. M. May and W. J. Leonard, "Nonlinear aspects of competition between three species," *SIAM J. Appl. Math.*, **29** (1975), 243–253.

42. R. M. May and G. F. Oster, "Bifurcations and dynamic complexity in simple ecological models," **110** (1975), 573–599.

43. R. M. May, G. R. Conway, M. P. Hassell, and T. R. E. Southwood, "Time delays, density dependence, and single species oscillations," *J. Anim. Ecol.*, **43** (1974), 747–770.

44. R. E. McMurtrie, "Stability of communities supported by depletable resources," 1975 (in press).

45. A. J. Nicholson and V. A. Bailey, "The balance of animal populations," Part I, *Proc. Zool. Soc. Lond.*, 1935, 551–598.

46. E. R. Pianka, *Evolutionary Ecology*, Harper and Row, New York, 1974.

47. E. C. Pielou, *Population and Community Ecology*, Gordon and Breach, New York, 1974.

48. F. A. Pitelka, "Some characteristics of microtine cycles in the arctic," *Arctic Biology*, (2nd ed., ed. H. P. Hansen), Oregon State University Press, Corvallis, 1967, 153–184.

49. R. A. Poluektov (ed.), *Dynamical Theory of Biological Populations*, Science Pubs., Moscow, 1974, (in Russian).

50. S. Rocklin, G. F. Oster, and A. Ipaktchi, "Phenotypic structure and bifurcation behavior in population models," *Theor. Pop. Biol.*, **10** (1976), 365–382.

51. D. Rogers and S. Hubbard, "How the behavior of parasites and predators promotes population stability," *Ecological Stability*, (eds: M. B. Usher and M. H. Williamson), Chapman and Hall, London, 1974, 99–122.

52. M. L. Rosenzweig, "Paradox of enrichment: destabilization of exploitation ecosystems in ecological time," *Science*, **171** (1971), 385–387.

53. J. Roughgarden, "The evolution of niche width," *Amer. Natur.*, **106** (1972), 683–718.

54. ———, "Population dynamics in a spatially varying environment: how population size "tracks" spatial variation in carrying capacity," *Amer. Natur.*, **108** (1974), 649–664.

55. T. W. Schoener, "Population growth regulated by intraspecific competition for energy or time: some simple representations," *Theor. Pop. Biol.*, **4** (1973), 56–84.

56. ———, "Resource partitioning in ecological communities," *Science*, **185** (1974), 27–39.

57. V. E. Shelford, "The abundance of the collared lemming in the Churchill area, 1929 to 1970," *Ecology*, **24** (1943), 472–484.

58. J. G. Skellam, "Random dispersal in theoretical populations," *Biometrika*, **38** (1951), 196–218.

59. M. Slatkin, "Selection and polygenic characters," *Proc. Nat. Acad. Sci.*, **66** (1970), 87–93.

60. T. R. E. Southwood, R. M. May, M. P. Hassell, and G. R. Conway, "Ecological strategies and population parameters," *Amer. Nat.*, **108** (1974), 791–804.

61. J. T. Tanner, "The stability and the intrinsic growth rates of prey and predator populations," *Ecology*, **56** (1975), 855–867.

62. G. C. Varley, G. R. Gradwell, and M. P. Hassell, *Insect Population Ecology*, Blackwell, Oxford, 1974.

63. A. R. Wallace, *Tropical Nature and Other Essays*, Macmillan, London, 1878.

64. P. J. Wangersky and W. J. Cunningham, "Time lag in population models," *Cold Spring Harb. Symp. Quant. Biol.*, **22** (1957), 329–338.

65. G. White, *The Natural History and Antiquities of Selbourne*, B. White, London, 1789.

66. R. H. Whittaker, *Communities and Ecosystems*, second ed., Macmillan, New York, 1975.

67. J. A. Yorke, "Limiting similarity among coexisting competitors," preprint 1975.

COMPARISON OF SOME DETERMINISTIC AND STOCHASTIC POPULATION THEORIES

Donald Ludwig

1. INTRODUCTION

The sizes of natural populations often exhibit fluctuations, with no apparent cause. Experiments which are repeated under what seem to be identical conditions may have completely different outcomes. It is tempting to attribute such fluctuations or differences to random effects. In the hope that regularities in behavior will emerge, attention is then concentrated upon ensembles of populations, or long series of observations.

The situation is illustrated by some classic experiments performed on competing species of flour beetles by Thomas Park (1948, 1954). Colonies consisting of members of two species were started in vials containing flour. The flour was renewed periodically, so that the beetles never died out. However, the population of one species or the other would always die out. The outcome of the competition depended upon such factors as the temperature and humidity. In certain borderline cases, the outcome would also depend upon the initial numbers of individuals which belonged to the two species. One may represent the pair of initial numbers (n_1, n_2) by a point in the plane. Thus Park and his coworkers set out to divide the first quadrant of the (n_1, n_2) plane into a region where the first species would win the competition, and a region where the second species would win. Imagine their surprise when it turned out that, for certain starting conditions, sometimes one species would win, and sometimes the other. This difficulty was resolved by Neyman, Park and Scott (1956) by introducing a stochastic model for the competition. The final result of the theory was the assignment of a probability $p(n_1, n_2)$ that the first species would win, given the starting values (n_1, n_2). The regularities which the experimenters sought could be found in the behavior of $p(n_1, n_2)$ as a function of n_1, n_2 and the environment, but not always in the results of single experiments.

Similar devices are often employed nowadays in the interpretation of biological data. Although stochastic theories seem to offer more realism than deterministic theories, they present a number of technical difficulties. Inevitably, there are more variables involved in the results and equations. For example, in the case of the flour beetles, the deterministic result would consist in dividing the n_1, n_2 plane into two regions, i. e., specifying the curve which forms their common boundary. The stochastic result consists in drawing all of the contours of equal probability. A deterministic model might consist of a set of two ordinary differential equations, or difference equations. A corresponding stochastic model (as we shall see) might consist of a large set of ordinary differential or difference equations, or perhaps a diffusion equation. One might expect either an exact solution or an easily computable numerical solution for a deterministic system. Such solutions are hardly ever available for stochastic models.

Such technical difficulties can be mitigated, if the qualitative behavior of a stochastic model can be given in terms of the qualitative behavior of the corresponding deterministic model. This procedure is especially appropriate in a biological context, since data are seldom sufficiently numerous or regular to permit fine distinctions to be made. Such qualitative results are illustrated below for three cases. They represent the colonization of an infinite region, colonization of an island, and spread of an epidemic. A more detailed account of these problems is given in Ludwig (1974). Excellent general references for stochastic population models are Karlin (1969) and Bartlett (1960).

2. COLONIZATION OF AN INFINITE REGION

By a colony, here we mean a population which starts at a size which is far below the carrying capacity of its environment. That is, there are no crowding effects such as shortages of shelter or food. The early European settlements in the Western Hemisphere are good examples of such colonies. Because of their use of a combination of military power, deceit and disease, the colonies encountered very little resistance from the native populations. Immigration was relatively unimportant in increasing the size of the population. After correcting for the effect of immigration, the total population of the colonies increased at a geometrical rate, doubling its size approximately every 25 years. Malthus was led to his famous Essay on the Principle of Population (1798, 1830) by considering this phenomenon.

A simple model which reproduces this behavior is

$$\frac{dx}{dt} = \lambda x - \mu x. \tag{2.1}$$

Here x represents the population size at time t, λ is the crude birth rate, and μ is the crude death rate. It follows from (2.1) that

$$x(t) = x(0)e^{(\lambda-\mu)t}. \tag{2.2}$$

Such a result may be quite satisfactory, if the age distribution of the population does not change much, and if crowding effects are not important. However, in the very early stages when the population is small, it may be more appropriate to describe the population size by an integer, and to take random effects into account. In fact, a number of the smaller colonies in America did not survive the initial stages. Thus (2.2) does not apply to them.

In the stochastic formulation let $X(t)$ represent the population size at time t. Let $p_j(t)$ be the probability that $X(t) = j$. We assume that, over a short time interval $(t, t + \delta t)$, the probability of a birth is $X(t)\lambda\, \delta t$, and the probability of a death is $X(t)\,\mu\, \delta t$. Thus λ and μ retain their interpretations as birth and death rates, respectively. Now we compute $p_j(t + \delta t)$. There are just three ways by which $X(t + \delta t) = j$. It could be that $X(t) = j$, and neither a birth nor a death occurs in the interval $(t, t + \delta t)$. The probability of this event is $p_j(t)(1 - j(\lambda + \mu)\delta t)$. Or, it could be that $X(t) = j - 1$ and a birth occurs in the interval. The probability of this event is $p_{j-1}(t)(j - 1)\lambda\, \delta t$. Or, it could be that $X(t) = j + 1$, and a death occurs in the interval. The third probability is $p_{j+1}(t)(j + 1)\,\mu\, \delta t$. It follows that

$$p_j(t + \delta t) = p_j(t)(1 - j(\lambda + \mu)\delta t) + p_{j-1}(j - 1)\lambda\, \delta t + p_{j+1}(j + 1)\,\mu\, \delta t. \tag{2.3}$$

Actually, (2.3) is an approximation which is only valid for small δt, since we have not considered the possibility of one birth and one death in the interval, etc. After subtracting $p_j(t)$ from both sides of (2.3), dividing by δt and passing to the limit as $\delta t \to 0$, the result is

$$\frac{d}{dt}\, p_j = (j - 1)\lambda p_{j-1} + (j + 1)\lambda p_{j+1} - j(\lambda + \mu)\,p_j, \quad j = 0, 1, \ldots . \tag{2.4}$$

These are the so-called forward equations for the linear birth and death process (see Karlin (1969), p. 193).

As a check on this procedure, we compute $p_0 + p_1 + p_2 + \cdots$, which should equal 1. If all of the equations (2.4) are added, the result is

$$\frac{d}{dt} \sum_{j=0}^{\infty} p_j(t) = \lambda \sum_{j=0}^{\infty} (j - 1)\,p_{j-1} + \mu \sum_{j=0}^{\infty} (j + 1)\,p_{j+1} - (\lambda + \mu) \sum_{j=0}^{\infty} jp_j. \tag{2.5}$$

By a change of summation variables, it can be seen that all three sums on the right-hand side of (2.5) are the same. Hence

$$\frac{d}{dt}\sum_{j=0}^{\infty} p_j(t) = 0. \quad \text{i.e.,} \sum_{j=0}^{\infty} \mathrm{p_j(t)} = \text{constant}. \tag{2.6}$$

The mean or expected population size is the sum which appeared three times on the right-hand side of (2.5). It can be computed by a trick similar to the previous one. If equations (2.4) are multiplied by j before adding, the result is

$$\frac{d}{dt}\sum_{j=0}^{\infty} jp_j = \lambda \sum_{j=0}^{\infty} j(j-1)p_{j-1} + \mu \sum_{j=0}^{\infty} j(j+1)p_{j+1}$$

$$- (\lambda + \mu)\sum_{j=0}^{\infty} j^2 p_j. \tag{2.7}$$

Now the three sums on the right-hand side of (2.7) are not identical. But they are nearly identical. This motivates the alternate form

$$\frac{d}{dt}\sum_{j=0}^{\infty} jp_j = \lambda \sum_{j=0}^{\infty} (j-1+1)(j-1)p_{j-1}$$

$$+ \mu \sum_{j=0}^{\infty} (j+1-1)(j+1)p_{j+1} - (\lambda + \mu)\sum_{j=0}^{\infty} j^2 p_j, \tag{2.8}$$

which can also be written as

$$\frac{d}{dt}\sum_{j=0}^{\infty} jp_j = \lambda \sum_{j=0}^{\infty} (j-1)p_{j-1} - \mu \sum_{j=0}^{\infty} (j+1)p_{j+1}$$

$$+\lambda \sum_{j=0}^{\infty} (j-1)^2 p_{j-1} + \mu \sum_{j=0}^{\infty} (j+1)^2 p_{j+1} - (\lambda + \mu)\sum_{j=0}^{\infty} j^2 p_j. \tag{2.9}$$

By changes in the summation variable, we see that the first two sums on the right-hand side of (2.9) are equal, and the last three sums there are also equal. Hence

$$\frac{d}{dt}\sum_{j=0}^{\infty} jp_j = (\lambda - \mu)\sum_{j=0}^{\infty} jp_j. \tag{2.10}$$

A comparison of (2.10) and (2.1) reveals that the mean population size obeys the corresponding deterministic equation.

Before leaping to the conclusion that the deterministic model (2.1) is an adequate substitute for (2.4), we should estimate the amount of spread in the distribution of $X(t)$. The variance or mean squared deviation of X is defined as

$$\sigma^2(t) = \sum_{j=0}^{\infty} (j - m(t))^2 p_j, \tag{2.11}$$

where

$$m(t) = \sum_{j=0}^{\infty} jp_j. \tag{2.12}$$

It follows from (2.11) and (2.12) that

$$\sigma^2(t) = \sum_{j=0}^{\infty} j^2 p_j - (m(t))^2. \tag{2.13}$$

If equations (2.4) are multiplied by j^2 before adding, the result can be manipulated into the form

$$\frac{d}{dt}\sigma^2 = (\lambda + \mu)m + 2(\lambda - \mu)\sigma^2. \tag{2.14}$$

For our purposes, it is better to calculate σ^2/m^2, rather than σ^2. The former quantity gives a measure of the spread of the distribution of X, relative to its mean value. If such a spread is small, one may use (2.1) instead of (2.4), without much loss of accuracy. It follows from (2.14), (2.10) and (2.12) that

$$\frac{d}{dt}(\sigma^2/m^2) = \frac{\lambda + \mu}{m}. \tag{2.15}$$

We shall assume that $X(0) = a$. Thus $m(0) = a$, and $\sigma^2(0) = 0$. Hence, from (2.10),

$$m(t) = a\, e^{(\lambda-\mu)t}. \tag{2.16}$$

After substitution of (2.16), (2.15) becomes

$$\frac{d}{dt}(\sigma^2/m^2) = \frac{\lambda+\mu}{a}\, e^{-(\lambda+\mu)t}. \tag{2.17}$$

Integration of (2.17) yields

$$\sigma^2/m^2 = \frac{1}{a}\,\frac{\lambda+\mu}{\lambda-\mu}\left(1 - e^{-(\lambda-\mu)t}\right). \tag{2.18}$$

If $\lambda > \mu$, (2.18) implies that σ^2/m^2 has the limiting value

$$\frac{\sigma^2(\infty)}{m^2(\infty)} = \frac{1}{a}\,\frac{\lambda+\mu}{\lambda-\mu}\,. \tag{2.19}$$

Therefore, the deterministic result (2.2) is applicable only if $a(\lambda - \mu)$ is large. If $\lambda < \mu$, (2.18) implies that the relative spread goes to infinity. In fact, σ^2 decreases at the same rate as m if $\lambda < \mu$, so the deterministic approximation is not appropriate.

A quantity of obvious interest is the extinction probability $p_0(t)$. We shall be especially concerned with the case where $X(0) = 1$. One shouldn't take such models too literally. For present purposes, let us shift the interpretation a bit, count only females in the population, and assume that there are always enough males to go around.

The dependence of $p_0(t)$ on the initial value of X will be indicated by the notation $p_0^a(t)$. Actually, the initial value a may be interpreted as an exponent. The reason is that a population becomes extinct if and only if the subpopulations started by each of the a founding members becomes extinct. In this linear model, these extinctions are independent events, and therefore the probability of a common extinction is the product of the probabilities of the simpler events.

The forward equations (2.4) do not yield a convenient form for

$p_0(t)$, since p_0 is coupled to p_1, etc. Instead, we shall use one of the so-called backward equations (see Karlin (1969), p. 192). During the interval $(0, \delta t)$, either a birth, a death or neither may occur. The probabilities of these events are $\lambda\delta t$, $\mu\delta t$ and $1 - (\lambda + \mu)\delta t$, as before. Thus $p_0(t + \delta_t)$ is given by

$$p_0(t + \delta t) = \lambda\delta t\, p_0^2(t) + \mu\delta t\, p_0^0(t) + (1 - (\lambda + \mu)\delta t)\, p_0(t). \tag{2.20}$$

The quantities $p_0^i(t)$ appear in (2.20) because, if there are i individuals at time δt, the probability that the whole population will be extinct by time $t + \delta t$ is $p_0^i(t)$. It follows from (2.20) that p_0 satisfies the differential equation

$$\frac{d}{dt} p_0 = \mu - (\lambda + \mu)p_0 + \lambda p_0^2 = (\mu - \lambda p_0)(1 - p_0). \tag{2.21}$$

Let ρ denote μ/λ, and let us first assume that $\rho < 1$. If $0 < p_0 < \rho$, then p_0 increases with time, and if $\rho < p_0 < 1$, then p_0 decreases. Hence $p_0(t) \to \rho$ as $t \to \infty$. This fact can also be verified by an explicit solution of (2.21). Similarly, if $\rho \geqslant 1$, it follows that $p_0(t) \to 1$ as $t \to \infty$. More generally, if $X(0) = a$, then the probability of eventual extinction is ρ^a if $\rho < 1$, and it is 1 if $\rho \geqslant 1$. The phenomenon of extinction may be an important correction to the deterministic theory if a is small, or if ρ is close to 1. Note that if ρ is close to 1, then

$$\rho^a = e^{a \log \rho} \sim e^{-a(1-\rho)}. \tag{2.22}$$

We conclude that the probability of extinction is negligible only if $a(1 - \rho)$ is large. This is exactly the same condition as was obtained from (2.19).

To summarize these results: if $\rho < 1$, then a proportion ρ^a of the colonies which start out with a individuals may be expected to die out. The populations of the remaining colonies increase roughly exponentially. However, the spread in their sizes may be substantial, unless a is large. More precise and more general statements to this effect are contained in the limit theorems for branching processes. See Karlin (1969), pp. 313–318, or Harris (1963), or Athreya and Ney (1972). The deterministic theory is adequate only if $a(1-\rho)$ is large.

3. COLONIZATION OF AN ISLAND

The preceding theory places no restraint upon the size of populations. Since the amounts of space and resources are always finite, no matter how large they may be, such a theory cannot be valid in case of very large populations. Since linear equations such as (2.1) always have exponential solutions, we must use nonlinear equations in order to model self-limiting populations. One feels that, as the population increases, if the birth rate does not decrease, then the death rate will certainly increase.

A simple equation which describes limited growth is the logistic equation

$$\frac{dx}{dt} = rx\left(1 - \frac{x}{K}\right). \tag{3.1}$$

The factor r corresponds to $\lambda - \mu$ in (2.1); it gives the net growth rate when x is small compared with K. It follows from (3.1) that x increases when $0 < x < K$, and x decreases if $x > K$. Therefore, $x \to K$ as $t \to \infty$. Hence K is called the carrying capacity of the environment.

In order to simplify the analysis of the corresponding stochastic problem, we shall not consider (3.1), but rather a chopped-off linear model:

$$\frac{dx}{dt} = \begin{cases} (\lambda - \mu)x & \text{if} \quad x < K, \\ 0 & \text{if} \quad x > K. \end{cases} \tag{3.2}$$

Solutions of (3.2) grow exponentially until they reach size K, and then remain at that value: if $x(0) = a$, then

$$x(t) = \begin{cases} ae^{(\lambda-\mu)t} & \text{if} \quad ae^{(\lambda-\mu)t} \leqslant K, \\ K & \text{if} \quad ae^{(\lambda-\mu)t} > K. \end{cases} \tag{3.3}$$

As a stochastic version of (3.2), we may employ (2.4) for $j = 0, 1, \ldots, K$, but modify the equation for p_{K+1}:

$$\frac{d}{dt} p_j = \lambda(j-1)p_{j-1} + \mu(j+1)p_{j+1} - (\lambda+\mu)jp_j, \quad j = 0, 1, \ldots, K, \tag{3.4}$$

$$\frac{d}{dt} p_{K+1} = (K-1)\lambda p_K - (K+1)\mu p_{K+1}. \tag{3.5}$$

These equations were proposed by MacArthur and Wilson (1967). They are also discussed in MacArthur (1972). The long term behavior of (3.4), (3.5) is completely different from (3.3). In order to see this, note that, if $X(t) \leqslant K$, then the probability that the next event is a death is equal to $\mu/(\lambda+\mu)$. If the population size is n, then the probability that there are no further births is $[\mu/(\lambda+\mu)]^n$. Since $n \leqslant K+1$, a uniform lower bound for the probability of no further births is $[\mu/(\lambda+\mu)]^{K+1}$. If K is large, this may be a small number. However, this argument shows that the probability of a fairly rapid extinction of the population is not zero, once K is fixed. Similarly, one can estimate the average length of time T required for such an extinction. If we wait for a period of length, say, $2T$, then a fraction at least $[\mu/(\lambda+\mu)]^{K+1}$ of the colonies will become extinct in that period. By repeating this argument, we see that eventually the proportion of colonies which become extinct approaches 1. In effect, if we are willing to wait long enough, an arbitrarily unlucky chain of events will occur, and the population will become extinct. Such a conclusion does not really depend upon any of our assumptions, except that the population sizes are bounded. The immediate and sweeping conclusion is that every bounded population is doomed to eventual extinction. This theoretical conclusion is well supported by the evidence: most populations which have ever existed are extinct.

Having faced up to the fact that all populations are going extinct, we ought to calculate how long they can expect to last. Such calculations were performed by MacArthur and Wilson (1967). Their motivation was as follows: ornithologists enjoy taking a census of bird populations on an island. It may be difficult to obtain an exact count, but a skilled observer can be fairly certain of finding at least one member of any species which is present. A number of islands were revisited after some years had passed. Frequently, some species which formerly were present could no longer be found, and other species had successfully colonized the island. Typically, the number of species present on the island would vary only slightly although many replacements had occurred. Moreover, the number of species found seemed to depend upon the area of the island, as well as factors which affect immigration. The obvious conclusion was that the islands were being colonized from time to time by new species, but that these colonies eventually became extinct. This suggests that one should attempt to estimate the time before extinction, in terms of the carrying capacity (area) of the island.

Let T_j denote the expected time to extinction, for a population which starts at size j. Let M_j denote the expected time for the first birth or death to occur. Then, in analogy with the derivation of (2.20), we have

$$T_j = M_j + \frac{\lambda}{\lambda+\mu} T_{j+1} + \frac{\mu}{\lambda+\mu} T_{j-1}, \quad (j \leqslant K). \tag{3.6}$$

That is, T_j is given by the expected time for a birth or death to occur, plus the expected elapsed time to extinction, after that event has occurred. The probability that the next event is a birth is $\lambda/(\lambda+\mu)$, and the probability that the next event is a death is $\mu/(\lambda+\mu)$.

In order to compute M_j, let $q_j(t)$ denote the probability that neither a birth nor a death occurs in the interval $(0, t)$. By the same reasoning as led to (2.4), q_j satisfies

$$\frac{dq_j}{dt} = -(\lambda+\mu) j q_j. \tag{3.7}$$

Hence

$$q_j(t) = e^{-j(\lambda+\mu)t}. \tag{3.8}$$

The expected time before the next event is given by

$$M_j = -\int_0^{\infty} t\, dq_j(t) = \frac{1}{j(\lambda + \mu)} \int_0^1 \log q\, dq = \frac{1}{j(\lambda + \mu)} . \quad (3.9)$$

If $j = K + 1$, instead of (3.6), we have

$$T_{K+1} = \frac{1}{\mu(K+1)} + T_K. \quad (3.10)$$

In order to solve (3.6) and (3.10), let

$$Z_j = T_j - T_{j-1}. \quad (3.11)$$

Details of the following manipulations are given in Ludwig (1974), pp. 18–19. If (3.6) is rewritten in terms of Z_j, the result is

$$Z_{j+1} = \rho Z_j + \frac{1}{\lambda j}, \quad j = 1, \ldots, K, \quad (3.12)$$

where $\rho = \mu/\lambda$, as before. The system (3.12) may be solved recursively, starting with $Z_1 = T_1$. The result is

$$Z_{j+1} = \rho^j T_1 - \frac{1}{\lambda} \sum_{l=1}^{j} \frac{1}{l} \rho^{K-j}, \quad j = 1, \ldots, K. \quad (3.13)$$

In addition, from (3.10), we have

$$Z_{K+1} = \frac{1}{\mu(K+1)} . \quad (3.14)$$

By comparing (3.14) and (3.13) (with $j = K$), we obtain a relation which can be solved for T_1:

$$T_1 = \frac{1}{\lambda} \sum_{j=1}^{K} \frac{1}{j} \rho^{-j} + \frac{\rho^{-K}}{\mu(K+1)} . \quad (3.15)$$

One could substitute this result into (3.13), in order to obtain formulas for the remaining T_j, for $j > 1$.

Although (or perhaps, because) the formula (3.15) gives T_1

exactly, it is not very useful. What is required is a simpler expression, which will enable us to estimate T_1 for large K. Such an estimate is given in Ludwig (1974), p. 19:

$$\lambda T_1 \sim \frac{1}{(K+1)(1-\rho)} e^{-(K+1)\log \rho}. \tag{3.16}$$

From this formula, it is evident that T_1 grows extremely rapidly as K increases, unless ρ is close to 1. For example, if $\rho = 1/2$ and $K = 10^6$, then λT_1 is greater than $10^{100,000}$, which is a good approximation to eternity. If ρ is close to 1, we may approximate (3.16) by

$$\lambda T_1 \sim \frac{1}{(K+1)(1-\rho)} e^{(K+1)(1-\rho)}. \tag{3.17}$$

We conclude that the population will persist a long time, if $(K+1)(1-\rho)$ is large. This condition might have been guessed from the results of the previous section. There we concluded that the deterministic result is valid if $1-\rho$ multiplied by the population size is large. In this case, the deterministic result is that the population remains at size K forever.

Another consequence of these asymptotic results is that, for large K,

$$T_1 \sim (1-\rho) T_K. \tag{3.18}$$

The derivation is given in MacArthur and Wilson (1967), p. 71, or Ludwig (1974), p. 17. This result has the following interpretation: a proportion ρ of the colonies which are started by a single individual become extinct before reaching size K, and their average survival time is very small. For colonies which reach size K, the expected time before extinction is very large. This result agrees completely with the results for the linear birth and death process.

4. A DETERMINISTIC EPIDEMIC MODEL

The following epidemic model was proposed by Kermack and McKendrick (1927). Every individual in the population belongs to one of three classes: susceptible, infective, and removed (immune, quarantined, dead, etc.). The infectious agent is spread through contact between infective and susceptible individuals. The rate of new infection is assumed to be proportional to the number of infectives, and to the number of susceptibles. If I, S, and R represent the numbers of infective, susceptible and removed individuals, respectively, then

$$\frac{dS}{dT} = -\beta IS, \tag{4.1}$$

since no infective or removed individuals revert to susceptible status. The rate at which infectives are removed is assumed to be proportional to the number of infectives:

$$\frac{dR}{dT} = \gamma I. \tag{4.2}$$

If N is the total population size, we must have $I + S + R = N$. Hence

$$\frac{dI}{dT} = \beta IS - \gamma I. \tag{4.3}$$

It is convenient to deal with scaled variables

$$i = I/N, \quad s = S/N, \quad r = R/N, \quad t = \beta NT, \quad \rho = \frac{\gamma}{\beta N}. \tag{4.4}$$

Then (4.1)–(4.3) become

$$\frac{ds}{dt} = -is, \tag{4.5}$$

$$\frac{dr}{dt} = \rho i, \tag{4.6}$$

$$\frac{di}{dt} = i(s - \rho). \tag{4.7}$$

Since the parameter ρ is the only one remaining, it must be crucial for the behavior of the epidemic. It is called the relative removal rate. From (4.7), it is apparent that i increases if $s > \rho$, and i decreases if $s < \rho$. Therefore, if the proportion of susceptibles in the population is maintained below ρ (for example, by immunization), then the introduction of a few infectives cannot cause an epidemic. This phenomenon is known as "herd immunity" (see J. P. Fox *et al.* (1971)). Unfortunately, the parameter ρ is difficult to estimate, and local concentrations of susceptibles may lead to epidemics. Although effective vaccines are available for many childhood diseases, they have not been eradicated from human populations. One reason is that there are local concentrations of children who do not receive adequate medical care.

If the initial value of s is above ρ, how many individuals will eventually be infected? That is, $s(\infty)$ should be determined in terms of $i(0)$ and $s(0)$. If (4.5) is divided by (4.6), the result is

$$\frac{ds}{dr} = -s/\rho, \tag{4.8}$$

or

$$\frac{ds}{s} + \frac{dr}{\rho} = 0. \tag{4.9}$$

The form (4.9) yields the integral

$$s\, e^{r/\rho} = \text{constant} = s(0)\, e^{r(0)/\rho}. \tag{4.10}$$

Since $r(\infty) = 1 - s(\infty)$ and $r(0) = 1 - s(0) - i(0)$, (4.10) implies that

$$s(\infty)e^{(1-s(\infty))/\rho} = s(0)e^{(1-s(0))/\rho}e^{-i(0)/\rho}. \tag{4.11}$$

In order to understand the behavior of $s(\infty)$, let us first neglect $i(0)$, and graph

$$f(s) = s\, e^{(1-s)/\rho}. \tag{4.12}$$

The function f has a maximum at $s = \rho$. If $\rho < 1$, this maximum is interior to the interval of interest: $0 < s < 1$. The function f is sketched in Fig. 1. If $s(0) > \rho$, then $s(\infty)$ is the point in the interval $(0, \rho)$ where $f(s(\infty)) = f(s(0))$. If $s(0) < \rho$, or if $\rho > 1$, then the apparent result is that $s(\infty) = s(0)$. Of course, the true value of $s(\infty)$ satisfies (4.11). Therefore $s(\infty)$ will be slightly below our graphical solution. Nevertheless, $s(\infty)$ will differ only slightly from $s(0)$, unless $s(0) > \rho$.

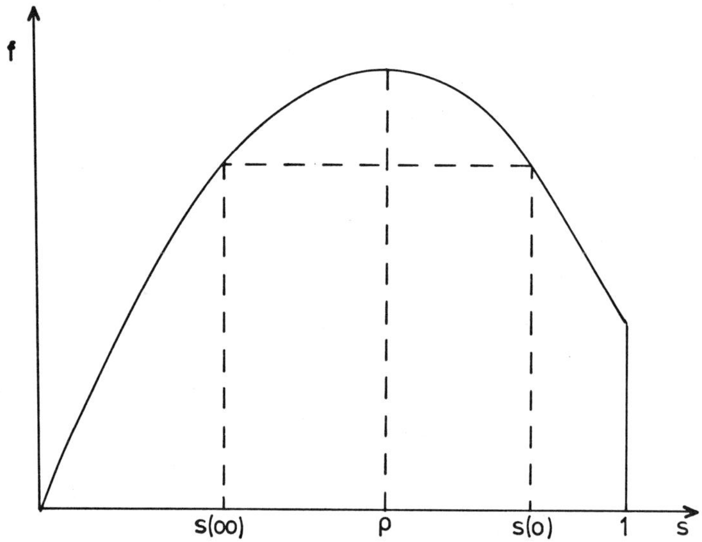

FIG. 1.

The result which was derived above is one of the "threshold theorems" of epidemic theory. An excellent account of generalizations and more recent developments is given in Waltman (1974).

5. A STOCHASTIC EPIDEMIC MODEL

A stochastic analogue of the Kermack-McKendrick model is known as the "general stochastic epidemic" (see Bailey (1957)). As before, let I, S and R denote the numbers of infective, susceptible and removed individuals, respectively. We assume that the probability of an infectious contact in the interval $(T, T + \delta T)$ is $\beta IS\ \delta T$, and the probability of a removal during that interval is $\gamma I\ \delta T$. If $p_{I,R}(t)$ denotes the probability of having I infectives, and R removed individuals at time t, it follows as in the derivation of (2.4) that

$$\frac{d}{dT} p_{I,R} = \beta(I-1)(N-I+1-R)p_{I-1,R} + \gamma(I+1)p_{I+1,R} - (\beta I(N-I-R) + \gamma I)p_{I,R}. \tag{5.1}$$

As in the previous case, let $t = \beta NT$, and $\rho = \gamma/\beta N$. Then (5.1) becomes

$$\frac{d}{dt} p_{I,R} = (I-1)\left(\frac{N-I+1-R}{N}\right)p_{I-1,R} + \rho(I+1)p_{I+1,R-1} - \left(I\left(\frac{N-I-R}{N}\right) + \rho I\right)p_{I,R}. \tag{5.2}$$

One cannot hope to describe the solution of the system (5.2) as easily as the solution of (4.5)–(4.7). However, for the stochastic models for population growth, the early stages of the process were crucial in determining the final outcome. Therefore, we may try to simplify (5.2), in order to describe the early stages of the epidemic. In the early stages, the numbers of both infectives and removals are small compared with N. Therefore (5.2) may be approximated by the system

$$\frac{d}{dt} p_{I,R} = (I-1)p_{I-1,R} + \rho(I+1)p_{I+1,R-1} - I(1+\rho)p_{I,R}. \tag{5.3}$$

The system (5.3) is remarkably similar to (2.4). In fact, if (5.3) is summed over R, the result is

$$\frac{d}{dt} p_I = (I-1)p_{I-1} + \rho(I+1)p_{I+1} - I(1+\rho)p_I, \tag{5.4}$$

where

$$p_I = \sum_R p_{I,R}. \tag{5.5}$$

The system (5.4) is identical with (2.4), if we set $\lambda = 1$ (through a scaling of time), and thus $\mu = \rho$.

The results of Section 2 suggest that, if $I = a$ and $R = 0$ initially, and if $\rho < 1$, then the probability that the epidemic will end quickly is ρ^a. Epidemics which do not end quickly will show a roughly exponential increase in the number of infectives, in the early stages. This idea was suggested by Whittle (1955) and Kendall (1956). The deterministic theory may be applied to epidemics which do not end quickly. In effect, the stochastic theory implies a modification of the Kermack-McKendrick theory. The final proportion of susceptibles $s(\infty)$ will be observed only in a proportion $1 - \rho^a$ of the cases in which a infectives are introduced. In the remaining proportion ρ^a of the cases, $s(\infty)$ will be close to $s(0)$.

If the final size probabilities $p_{O,R}(\infty)$ are plotted, one expects a U-shaped distribution, with area ρ^a under the left-hand branch, and with area $1 - \rho^a$ under the right-hand branch, centered at $Nr(\infty) = N(1 - s(\infty))$. Such a distribution is sketched as the solid line in Figure 2. The corresponding exact result (computed by a laborious numerical method) is shown in the dotted curve on the same figure. There is a much larger spread in the right-hand hump than one would expect from a deterministic theory. One might explain this discrepancy as follows: in the later stages of a large epidemic, the number of infectives again falls to low levels. In such a case, one cannot replace the system (5.3) by the Kermack-McKendrick equations.

One way to improve upon the deterministic approximation is to employ a diffusion equation. The complete analysis of such equations leads to many technical difficulties and unresolved mathematical questions. We shall merely indicate how a diffusion equation may be derived from our stochastic model. The simplest approach is to regard the system (5.1) as a set of difference

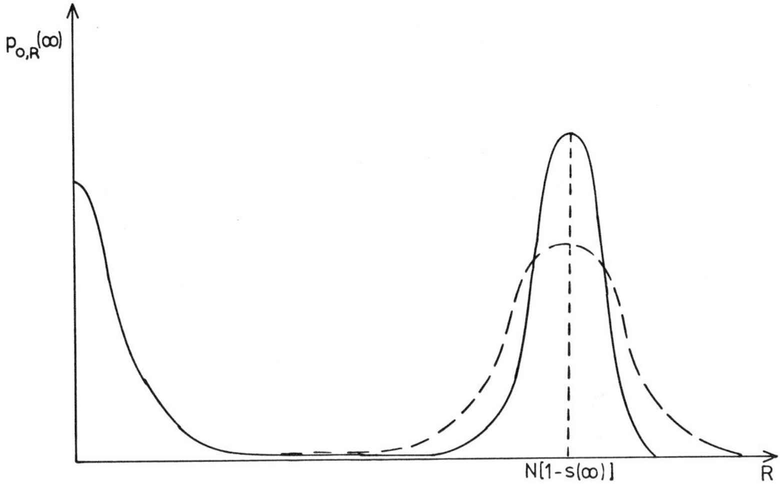

FIG. 2.

equations, and to approximate the differences by Taylor expansions. The most convenient variables to employ are R and S. Let $P_{S,R}(t)$ denote the probability of having S susceptibles and R removals at time t. The analogue of (5.2) is

$$\frac{d}{dt} P_{S,R} = (S+1)\left(\frac{N-S-1-R}{N}\right)P_{S+1,R} + (N-S-R+1)\rho P_{S,R-1} - (N-S-R)\left(\frac{S}{N}+\rho\right)P_{S,R}. \tag{5.6}$$

If we set

$$s = \frac{S}{N}, \quad r = \frac{R}{N}, \quad v(s,r,t) = P_{S,R}(t), \tag{5.7}$$

then (5.6) may be rewritten as

$$\frac{\partial}{\partial t} v(s, r, t) = N\left(s + \frac{1}{N}\right)\left(1 - r - s - \frac{1}{N}\right)v\left(s + \frac{1}{N}, r, t\right)$$

$$- N s(1 - r - s)v(s, r, t) + N\rho\left(1 - s - r + \frac{1}{N}\right)$$

$$\times v\left(s, r - \frac{1}{N}, t\right) - N\rho(1 - s - r)v(s, r, t). \quad (5.8)$$

Let

$$f(r, s, t) = s(1 - r - s)v(s, r, t),$$

$$g(r, s, t) = \rho(1 - r - s)v(s, r, t). \quad (5.9)$$

Then (5.7) can be rewritten as

$$\frac{\partial}{\partial t} v(s, r, t) = N\left[f\left(s + \frac{1}{N}, r, t\right) - f(s, r, t)\right]$$

$$+ N\left[g\left(s, r - \frac{1}{N}, t\right) - g(s, r, t)\right]. \quad (5.10)$$

If the differences in f and g are approximated by Taylor expansions, the result is

$$\frac{\partial}{\partial t} v(s, r, t) = \frac{\partial f}{\partial s}(s, r, t) + \frac{1}{2N}\frac{\partial^2 f}{\partial s^2}(s, r, t) + \cdots$$

$$- \frac{\partial g}{\partial r}(s, r, t) + \frac{1}{2N}\frac{\partial^2 g}{\partial r^2}(s, r, t) + \cdots. \quad (5.11)$$

This expansion suggests the following diffusion approximation for

v:

$$\frac{\partial v}{\partial t} = \frac{\partial}{\partial s}(s(1-r-s)v) - \frac{\partial}{\partial r}(\rho(1-r-s)v) + \frac{1}{2N}\left[\frac{\partial^2}{\partial s^2}(s(1-r-s)v) + \frac{\partial^2}{\partial r^2}(\rho(1-r-s)v)\right]. \quad (5.12)$$

If the second derivative terms in (5.12) are dropped, the resulting first-order partial differential equation is equivalent to the Kermack-McKendrick system (4.4), (4.5) (see Ludwig (1974), pp. 45–49). Thus (5.12) represents a correction to the deterministic theory. A fairly straightforward but lengthy calculation yields a Gaussian (or normal) density v, which smears out the spike at $N(1 - s(\infty))$ in Figure 2. Details are given in Ludwig (1974) or Ludwig (1975).

6. CONCLUDING REMARKS

The preceding examples illustrate how diverse problems can lead to quite similar mathematical phenomena and theories. Many other examples could have been taken from population genetics (see Crow and Kimura (1970), and other articles in this volume). Nevertheless, there is no general mathematical theory for population biology, which is analogous to the classical theory of mathematical physics.

A general theory would, somehow, have to incorporate the qualitative theory of ordinary differential equations, limiting behavior of discrete stochastic processes, and a qualitative theory for diffusion equations. These three fields have developed in almost complete isolation from each other. We may look forward to exciting developments, as the necessary connections are made between these fields.

REFERENCES

1. K. B. Athreya and P. E. Ney, *Branching Processes*, Springer-Verlag, Berlin and New York, 1972.

2. N. T. J. Bailey, *The Mathematical Theory of Epidemics*, Griffin, London, 1957.

3. M. S. Bartlett, *Stochastic Population Models in Ecology and Epidemiology*, Methuen, London, 1960.
4. J. P. Fox, L. Elveback, W. Scott, L. Gatewood, and E. Ackerman, "Herd immunity: Basic concept and relevance to public health immunization practices," *Am. J. Epid.*, **94** (3) (1971), 179–189.
5. Th. E. Harris, *The Theory of Branching Processes*, Springer-Verlag, Berlin, 1963.
6. S. Karlin, *A First Course in Stochastic Processes*, Academic Press, New York, 1966.
7. D. G. Kendall, "Deterministic and stochastic epidemics in closed populations," *Proc. 3rd Berkeley Symp. Math. Stat. Prob.*, vol. VI (1956), 149–165.
8. W. O. Kermack and A. G. McKendrick, "A contribution to the mathematical theory of epidemics," *Proc. Roy. Soc. Lond. Ser. A*, **115** (1927), 200–221.
9. D. Ludwig, *Stochastic Population Theories*, Lecture Notes in Biomathematics, vol. 3, Springer-Verlag, Berlin and New York, 1974.
10. ———, "Qualitative behavior of stochastic epidemics," *Math. Biosci.*, **23** (1975), 47–73.
11. R. H. MacArthur, *Geographical Ecology*, Harper & Row, New York, 1972.
12. R. H. MacArthur and E. O. Wilson, *The Theory of Island Biogeography*, Princeton University Press, Princeton, 1967.
13. J. Neyman, T. Park, and E. Scott, "Struggle for existence. The Tribolium model: biological and statistical aspects," *Proc. 3rd Berkeley Symp. Math. Stat. Prob.*, vol. VI (1956), 41–79.
14. T. Park, "Experimental studies of interspecies competition. I. Competition between populations of the flour beetles Tribolium confusum Duval and Tribolium castaneum Herbst," *Ecol. Monographs*, **18** (1948), 265–308.
15. ———, "Experimental studies of interspecies competition. II. Temperature, humidity, and competition in two species of Tribolium," *Physiol. Zool.*, **27** (1954), 177–238.
16. P. Waltman, *Deterministic Threshold Models in the Theory of Epidemics*, Lecture Notes in Biomath., vol. 1, Springer-Verlag, Berlin and New York, 1974.
17. P. Whittle, "The outcome of a stochastic epidemic—a note on Bailey's paper," *Biometrika*, **42** (1955), 116–122.

AGE-STRUCTURED EQUATIONS IN THE THEORY OF CELL POPULATIONS

S. I. Rubinow

The entire subject of cell population theory appears to have assumed the most prevalent characteristic of its object of study: the number of papers in the subject have increased exponentially during the past twenty years or so. Here we shall largely confine our remarks to the age-structured equation in widest use that represents such populations, that due to Scherbaum and Rasch (1957) and von Foerster (1959). We review its solution both for short and long times, and derive a number of important consequences. Some of the latter have origins that predate such a derivation. Their derivation here, in a unified fashion, illustrates the comprehensive character of the theory. Finally, we discuss briefly other theories of cell populations that have been proposed. We omit, by and large, discussion of the many applications of these theories that have been made to experiments.

1. THE SCHERBAUM-RASCH-VON FOERSTER EQUATION

A microbial population grown in a culture in a constant environment will, after a transient period, achieve a *steady state of exponential growth*. In this state, the dependence on the time t of the total population $N(t)$ is expressed as the solution of the differential equation

$$\frac{dN}{dt} = \gamma N, \quad t > 0, \tag{1}$$

where γ is a positive constant, the *specific growth rate* or *fractional growth rate*. According to equation (1), the population undergoes Malthusian growth, that is to say,

$$N(t) = N_0 e^{\gamma t}, \tag{2}$$

where N_0 is the initial population at time zero. The constant environment requires that "crowding" effects are negligible.

When cells are grown by the batch culture technique, the environment is of fixed and finite extent, and can not be considered to be constant for an extended period of time. Thus, nutrients such as oxygen are depleted, the pH changes, and effects of crowding become established, for example, as a result of toxins released by growing cells. Such effects can be accounted for to some extent by assuming [Verhulst (1845, 1847)] that $N(t)$ satisfies the equation

$$\frac{dN}{dt} = kN\left(1 - N/\overline{N}\right), \tag{3}$$

which has as its solution the familiar logistic law

$$N = \frac{N_0 \overline{N}}{N_0 + \left(\overline{N} - N_0\right)e^{-kt}}. \tag{4}$$

At large times, N approaches the stable equilibrium value $\overline{N}$.

At the present time it is possible to observe experimentally in a quantitative manner other features of a cell population besides its number. For example, volume, mass, DNA content, the durations of the various phases of the cell cycle, the number of cells in mitosis, the number of cells in the DNA-synthesizing phase (so-called S-phase) of the cell cycle, and so forth. [See, for example, Nachtwey and Cameron, 1968.] To take advantage of observations which relate to age-dependent properties, age-structured theories of cell populations have been introduced. The idea of an age-dependent theory of populations appears already in the early demographic work of Lotka (1907).

The fundamental conception of the theory is the *age density function* $n(a, t)$, where a represents the age of a cell, the time elapsed since its birth. Here $n(a, t)\,da$ represents the total population number in the age interval a to $a + da$ at time t. Thus, the total population at time t is expressed in terms of it as

$$N(t) = \int_0^\infty n(a, t)\,da. \tag{5}$$

The age density function was first proposed by Scherbaum and Rasch (1957) to satisfy the simple partial differential equation

$$\frac{\partial n}{\partial t} + \frac{\partial n}{\partial a} = 0, \quad a > 0, t > 0. \tag{6}$$

Subsequently, the possibility of cell loss was taken into account by von Foerster who modified equation (6) to read

$$\frac{\partial n}{\partial t} + \frac{\partial n}{\partial a} = -\lambda n, \quad a > 0, t > 0, \tag{7}$$

where λ is the probability per unit time that a cell disappears from the population. Equation (7) is usually called the von Foerster equation, although it would seem more proper to call it the Scherbaum-Rasch-von Foerster equation, or more simply, the SRF equation, as we shall call it herein. Disappearance has two principal causes, cell mitosis, which is followed by cell division, and cell death. Hence it is convenient to express the two contributions due to cell death and mitosis separately as

$$\lambda = \lambda_d + \lambda_m. \tag{8}$$

In general, λ_m is expected to be age-dependent, and λ_d may be both age- and time-dependent, but is usually only age-dependent in experiments conducted with microbial cultures. For example, for cells grown by the continuous culture technique as in a chemostat, the population is maintained at a constant level by means of efflux of old broth (nutrient supply) and cells, while new broth is brought into the medium. The culture is stirred continuously. In such a case λ_d is a constant independent of age and time.

A unique solution to equation (7) requires the specification of the initial condition and a boundary condition. The initial condition is written as

$$n(a, 0) = \varphi(a), \quad a > 0, \tag{9}$$

where $\varphi(a)$ is the initially prescribed age distribution of the population. The boundary condition, which expresses how cells are born, is

$$n(0, t) = 2\int_0^\infty \lambda_m(\tau)n(\tau, t)\, d\tau, \quad t > 0. \tag{10}$$

The factor of two appears on the right hand side above because division of a cell of age τ in the population produces two cells of age zero.

2. SOLUTION BY GENERATIONS

We shall give the solution to equations (7)–(10) by means of the generation expansion [Rubinow, 1968]. First, we observe that a formal solution to equation (7) without regard to either initial or boundary conditions is given [Trucco, 1965] as

$$\chi(a - t)\exp\left[-\int_{\psi(a-t)}^{a} \lambda(\xi, t - a + \xi)\, d\xi\right], \tag{11}$$

where χ and ψ are arbitrary functions. Now let

$$n(a, t) = \sum_{j=1}^{\infty} n_j(a, t), \tag{12}$$

where $n_j(a, t)$ is the age-time generation density function associated with the jth generation. The function n_1, representing the initial cohort of cells or first generation, is chosen to satisfy the initial condition that

$$n_1(a, 0) = \varphi(a). \tag{13}$$

From (11), we see that by choosing $\chi(a) = \varphi(a)$ and $\psi(a - \mathrm{t})$ equal to $a - t$, the function $n_1(a, t)$ is given as

$$n_1(a, t) = \varphi(a - t)\exp\left[-\int_{a-t}^{a} \lambda(\xi, t - a + \xi)\, d\xi\right], \quad 0 < t < a. \tag{14}$$

The cells in the first generation give birth to cells of the second generation in accordance with the birth condition (10). Hence

$$n_2(0, t) = 2\int_0^\infty \lambda_m(\tau) n_1(\tau, t)\, d\tau. \tag{15}$$

We observe here that the number of cells in the age interval a to $a + da$ at time t is the same as the number of cells that were born at a time a earlier, times the probability that those cells survived to age a. In terms of the cell density function, this relation reads

$$n(a, t) = n(0, t - a)\exp\left[-\int_0^a \lambda(\xi, t - a + \xi)\, d\xi\right], \quad t > a. \tag{16}$$

By combining equations (15) and (16), we see that $n_2(a, t)$ is given as

$$n_2(a, t) = \begin{cases} 0, \quad t < a, \\ 2\displaystyle\int_{t-a}^{\infty} \lambda_m(\tau)\varphi(\tau + a - t) \\ \cdot\exp\left[-\displaystyle\int_{\tau+a-t}^{\tau} \lambda(\xi, t - a - \tau + \xi)\, d\xi \right. \\ \left. -\displaystyle\int_0^a \lambda(\xi, t - a + \xi)\, d\xi\right] d\tau, \; t > a. \end{cases} \tag{17}$$

Here, the lower limit in the integral on the right has been changed from 0 to $t - a$ because $\varphi\ (\tau + a - t)$ is zero when $\tau < t - a$.

By continuing in the same manner for succeeding generations, the jth generation is obtained recursively as follows:

$$n_j(0, t) = \int_0^{\infty} \lambda_m(\tau) n_{j-1}(\tau, t)\, d\tau = 2\int_0^{t} \lambda_m(\tau) n_{j-1}(\tau, t)\, d\tau, \quad j = 3, 4, \ldots . \tag{18}$$

The last equality above follows because $n_{j-1}(\tau, t)$ equals zero unless $\tau < t$, for all $j > 3$. In the same way that $n_2(a, t)$ was derived from $n_2(0, t)$ we deduce from (18) that

$$n_j(a, t) = \begin{cases} 0, & t < a, \\ 2\int_0^{t-a} \lambda_m(\tau) n_{j-1}(\tau, t - a) \\ \cdot \exp\left[-\int_0^{a} \lambda(\xi, t - a + \xi)\, d\xi \right] d\tau, & t > a, \end{cases} \tag{19}$$

$$j = 3, 4, \ldots .$$

When λ is a function of age alone, which is usually the case, equations (14), (17), and (19) simplify with the aid of equation (16) to

$$n_1(a, t) = \varphi(a - t)\exp\left[-\int_{a-t}^{a} \lambda(\xi)\, d\xi \right], \quad 0 < t < a,$$

$$n_2(a, t) = 2\int_{t-a}^{\infty} n_1(a, t - \tau) \lambda_m(\tau) \exp\left[-\int_0^{\tau} \lambda(\xi)\, d\xi \right] d\tau, \quad t > a,$$

$$n_j(a, t) = 2\int_0^{t-a} n_{j-1}(a, t - \tau) \lambda_m(\tau) \exp\left[-\int_0^{\tau} \lambda(\xi)\, d\xi \right] d\tau,$$

$$t > a, \quad j = 3, 4, \ldots . \tag{20}$$

The total population at any time is given by substituting equation (12) into equation (5) and utilizing equations (14), (17), and

(19). The generation expansion can be expected to be useful during the early time of growth of a cell population, when at most a few generations contribute to the sum.

3. SOLUTION BY MEANS OF AN INTEGRAL EQUATION

As in equation (14), the function

$$n(a, t) = \varphi(a - t)\exp\left[-\int_{a-t}^{a} \lambda(\xi, t - a + \xi)\, d\xi\right], \quad 0 < t < a, \tag{21}$$

representing those cells of the given initial age distribution surviving to time t that have not yet divided, satisfies the differential equation (7) and the initial condition (9). The solution expressed by means of equations (16) and (21) was first given by Trucco (1965). For equation (16) to be useful, however, it is necessary to know how to obtain $n(0, t)$, the birth rate at time t. To this end, substitute for $n(\tau, t)$ on the right-hand side of equation (10) the expressions given by (16) and (21). Thus [Trucco, 1965],

$$\begin{aligned} n(0, t) = {} & 2\int_0^t \lambda_m(\tau) n(0, t - \tau)\exp\left[-\int_0^\tau \lambda(\xi, t - \tau + \xi)\, d\xi\right] d\tau \\ & + 2\int_t^\infty \lambda_m(\tau)\varphi(\tau - t)\exp\left[-\int_{\tau - t}^{\tau} \lambda(\xi, t - \tau + \xi)\, d\xi\right] d\tau. \end{aligned} \tag{22}$$

This equation is an inhomogeneous integral equation for the birth rate, and is perhaps more recognizable as such if we change the variable of integration in the first integral on the right-hand side by setting $\tau = t - \eta$, and for convenience set $\tau = t + \eta$ in the second integral, e.g.,

$$\begin{aligned} n(0, t) = {} & 2\int_0^t \lambda_m(t - \eta) n(0, \eta)\exp\left[-\int_0^{t-\eta} \lambda(\xi, \eta + \xi)\, d\xi\right] d\eta \\ & + 2\int_0^\infty \lambda_m(\eta + t)\varphi(\eta)\exp\left[-\int_{\eta}^{\eta + t} \lambda(\xi, - \eta + \xi)\, d\xi\right] d\eta. \end{aligned} \tag{23}$$

From equation (22), we also know $n(0, t - a)$, $t > a$. Hence, by again making use of equation (16), we obtain the formal solution to equations (7)–(10) as

$$n(a,t) = \begin{cases} \varphi(a-t)\exp\left[-\int_{a-t}^{a} \lambda(\xi, t-a+\xi)\, d\xi\right], & t < a, \\ 2\int_0^{t-a} \lambda_m(\tau) n(0, t-a-\tau)\exp\left[-\int_0^{\tau} \lambda(\xi, t-a-\tau+\xi)\, d\xi - \int_0^{a} \lambda(\xi, t-a+\xi)\, d\xi\right] d\tau & \\ +2\int_{t-a}^{\infty} \lambda_m(\tau)\varphi(\tau+a-t)\exp\left[-\int_{\tau+a-t}^{\tau} \lambda(\xi, t-a-\tau+\xi)\, d\xi - \int_0^{a} \lambda(\xi, t-a+\xi)\, d\xi\right] d\tau, & \\ t > a & \end{cases} \tag{24}$$

The solution expressed in this manner is seen to depend on the solution to an integral equation for the birth rate, equation (22). Moreover, the birth rate is a function of only one variable, the time t.

4. DETERMINATION OF THE FUNCTION λ_m

How is the function λ_m determined? Assume a cell population is being grown in a culture in a state of steady exponential growth, and that no cell deaths are taking place. Then λ and λ_m are identical. Imagine an experiment, biologically realizable, in which a synchronized subpopulation of N_0 cells, all of age zero at time zero, are observed, and the fraction of cells that subsequently undergo cell division in the time interval τ to $\tau + d\tau$ is recorded. We let this number equal $g(\tau)\, d\tau$, and the function $g(\tau)$ determined in this manner is called the *generation time distribution function*. The generation time distribution function, related functions, and the relationships among the generations of daughter, sister, and mother cells have been extensively studied for bacterial populations. [See, for example, Kelly and Rahn (1932); Powell (1955, 1956, 1958, 1959); Hughes (1955); Powell and Errington (1963); Prescott (1959); Koch and Schaechter (1962); Schaechter *et al.* (1962); Kubitschek (1962, 1966a, b, 1967, 1971); Harvey (1972).] The function is generally found to be Gaussian in nature and skewed in the direction of long generation times.

From the above definition, we see that $g(\tau)$ is equivalent to the expression

$$g(\tau) = \frac{n_2(0, \tau)}{\int_0^\infty n_2(0, \tau)\, d\tau}, \tag{25}$$

where $n_2(0, t)$ is given by equation (15), and $n_1(a, t)$ is given by equation (14) with $\lambda_d = 0$, λ_m a function of age alone, and

$$\varphi(a) = N_0\, \delta(a). \tag{26}$$

Here $\delta(a)$ is the Dirac delta function. By direct substitution, we find from equation (14) that in this case

$$n_1(a, t) = N_0\, \delta(a - t)\exp\left[-\int_{a-t}^{a} \lambda_m(\xi)\, d\xi\right], \quad 0 < t < a. \tag{27}$$

We note from equation (25) that $g(\tau)$ is normalized so that

$$\int_0^\infty g(\tau)\, d\tau = 1. \tag{28}$$

Thus, from equations (15) and (25),

$$g(\tau) = \lambda_m(\tau)\exp\left[-\int_0^\tau \lambda_m(\xi)\, d\xi\right]. \tag{29}$$

By integrating both sides of this expression with respect to τ, between the limits 0 and τ, we find that

$$\int_\tau^\infty g(\xi)\, d\xi = \exp\left[-\int_0^\tau \lambda_m(\xi)\, d\xi\right]. \tag{30}$$

The integral on the left represents the fraction of the population with a generation time $\geqslant \tau$. Hence, from (29) and (30), we determine λ_m from $g(\tau)$ as

$$\lambda_m(\tau) = \frac{g(\tau)}{\int_\tau^\infty g(\xi)\, d\xi}. \tag{31}$$

As an incidental remark, we note that the mean interdivision time $\bar{\tau}$ defined as $\int_0^\infty \tau g(\tau)\, d\tau$ is not the same as the doubling time $\log 2/\gamma$ of the steady-state population (Powell, 1956). In fact, it can be shown [Painter and Marr, 1967; see also Engelberg (1964), Barrett (1966)] that $\bar{\tau} \approx \log 2/\gamma + \sigma^2\gamma/2$, where σ is the standard deviation of $g(\tau)$.

5. ASYMPTOTIC SOLUTION FOR LONG TIMES

Equation (24) is useful for examining the behavior of the solution when t is large. For a population of cells that is growing or at least maintaining itself, we expect a steady state of exponential growth to be achieved, for which $n(a, t)$ will have the asymptotic form

$$n(a, t) \sim e^{\gamma t}\bar{n}(a), \quad t \to \infty. \tag{32}$$

Here, $\bar{n}(a)$ represents the steady-state age density function of the population. Nooney (1967) has shown that the above limiting form of n is attained for large times, independent of the initial conditions.

Let us specialize equation (24) to the case when $\lambda = \lambda(a)$, a function of age alone. By substituting (32) into (24), dividing by $e^{\gamma t}$, and letting $t \to \infty$, we obtain the relation

$$\bar{n}(a) = 2\bar{n}(0)\exp\left[-\gamma a - \int_0^a \lambda(\xi)\, d\xi\right]\int_0^\infty \lambda_m(\tau) \cdot\exp\left[-\gamma\tau - \int_0^\tau \lambda(\xi)\, d\xi\right] d\tau. \tag{33}$$

By setting $a = 0$ in equation (33), and using (29), we infer directly that the last integral on the right above equals 1/2, or

$$1 = 2\int_0^\infty \exp\left[-\gamma\tau - \int_0^\tau \lambda_d(\xi)\,d\xi\right] g(\tau)\,d\tau. \tag{34}$$

With λ_d equal to zero, this result was given by Powell (1956). However, essentially this same result was known already by Lotka (1907). It is attributable to Euler [see Wilson and Bossert (1971)]. Equation (34) can be viewed as a constraint of consistency on the intrinsic properties of the cell population, γ and $g(\tau)$. For example, if $\lambda_d = 0$, then equation (34) and the knowledge of $g(\tau)$ determine the fractional growth rate γ uniquely. Note that if the death rate function λ_d is too large, equation (34) cannot be satisfied for any positive value of γ, and a state of steady exponential growth with increasing population cannot be achieved.

With the aid of (30) and (34), equation (33) becomes

$$\bar{n}(a) = \bar{n}(0)\exp\left[-\gamma a - \int_0^a \lambda_d(\xi)\,d\xi\right]\int_a^\infty g(\tau)\,d\tau. \tag{35}$$

For the case $\lambda_d = 0$, this result was given by Powell (1956) and others [Painter and Marr (1968)] without the aid of the SRF equation. The function $\bar{n}(a)$ is another intrinsic property of a cell population, and we see from the above relation that it too is intimately related to $g(\tau)$. It is perhaps slightly more convenient to introduce [Lotka (1907); Powell (1956)] the *steady-state age frequency function* $\tilde{n}(a)$, which is just $\bar{n}(a)$ normalized so that

$$\int_0^\infty \tilde{n}(a)\,da = 1. \tag{36}$$

Thus, from (35),

$$\tilde{n}(a) = C\exp\left[-\gamma a - \int_0^a \lambda_d(\xi)\,d\xi\right]\int_a^\infty g(\tau)\,d\tau, \tag{37}$$

$$C^{-1} = \int_0^\infty \exp\left[-\gamma a - \int_0^a \lambda_d(\xi)\,d\xi\right]\int_a^\infty g(\tau)\,d\tau\,da. \tag{38}$$

When λ_d is a constant, C^{-1} equals $1/2\ (\gamma + \lambda_d)$, and $\tilde{n}(a)$ simplifies to [Powell, 1956]

$$\tilde{n}(a) = 2(\gamma + \lambda_d)\exp[-(\gamma + \lambda_d)a]\int_a^\infty g(\tau)\,d\tau. \tag{39}$$

We recognize that $\exp(-\lambda_d a)$ times the last integral above is just the probability that a member of the population survives to age a. Expressed in this manner, equation (39) was given by Lotka (1907).

6. OTHER AGE-STRUCTURED THEORIES

According to the SRF equation, each newly born cell has the same probability of subsequently undergoing division at a later time at age a, and this probability is described by the function $\lambda_m(a)$. In other words, the SRF equation assumes that the generation time is not determined at birth, but is an inherently probabilistic characteristic of a cell. Hence, the SRF equation assumes implicitly that, on the average, there is no correlation between the generation times of mothers and daughters. As mentioned previously, this relationship has been extensively investigated for various microbial populations, by direct observations of individual cellular generation times. Most of these studies suggest that the generation times of mothers and daughters are weakly correlated at best, or perhaps negatively correlated. The implications of such studies on the theory of the growth of an entire population is unclear at the present time.

A direct test of the SRF theory was provided by observations of Prescott (1959) of the growth in time of a population of *tetrahymena geleii* HS, a single-celled organism. He followed in time the colonies originating from 50 individual cells, for several generations. By combining these measurements as if they had a common time origin, the evolution in time of a "large" single population could be very accurately obtained. The generation time distribution function of these was also determined. Hence, λ_m could be inferred, and the SRF equation could be used [Rubinow,

1968] to predict the growth of the population. An alternative hypothesis that was used to compare the observations with theory is that the generation time of a daughter cell is exactly the same as that of its mother. In the latter view, the total population consists of a set of subpopulations, each with a pre-assigned generation time τ. Each subpopulation can be represented by a cell density function $n(a, t; \tau)$ that satisfies the SRF equation with $\lambda_m = 0$, in the domain $0 < a < \tau, t > 0$. The total population at any time t is given in terms of n as

$$N(t) = \int_0^\infty \int_0^\tau n(a, t; \tau)\, da\, d\tau. \tag{40}$$

Somewhat surprisingly, it was found that this latter theory provided a much better fit with Prescott's observations than did the SRF theory. This failure of the SRF theory to represent adequately Prescott's experiments is perhaps the most serious objection that has been raised to it.

The alternative theory described above suggests that memory of the generation time persists from one generation to the next, at least over a time span of several generations. This concept is supported by some other observations [Hughes (1955); Powell and Errington (1963)]. However, the serious limitation of this viewpoint is that it can not be true for very large times because, if it were, the cells with the shortest generation time would ultimately prevail, and the dispersed generation time distribution that is always observed would not be an invariant feature of microbial cell populations.

This limitation can be overcome [Lebowitz and Rubinow (1974)] by the assumption that the birth process of the cell population under consideration is characterized by a transition kernel $K(\tau, \tau')$, where $K(\tau, \tau')\, d\tau$ is the fraction of cells born from mothers with generation time τ', that will have a generation time between τ and $\tau + d\tau$. Thus, in this formulation of the theory the birth condition (10) is replaced by

$$n(0, t; \tau) = 2\int_0^\infty K(\tau, \tau')n(\tau', t; \tau')\, d\tau', \tag{41}$$

and the initial condition becomes

$$n(a, 0; \tau) = \varphi(a; \tau), \tag{42}$$

where φ is a prescribed function of a and τ. There is then no difficulty in satisfying the requirement that $g(\tau)$ be an invariant feature of the cell population. In the original reference can be found a derivation of results that are generalizations of the ones presented herein in terms of the SRF theory. The SRF theory is, in fact, a special case of the above theory for which $K(\tau, \tau') = k(\tau)$, a function independent of τ'. Unfortunately, the above theory demands for its usefulness the knowledge of the transition kernel K, which has not been determined for any cell population at the present time.

Rahn (1932) suggested that the division process consists of a large number of primitive biochemical steps, each step subject to probabilistic variation. This view of division as a Markoff process was further developed by Kendall (1948), and leads in a natural way to the kind of generation time distribution function that is observed. A more explicit linear chemical kinetic model of the control of cell division has been recently proposed by Woolley and De Rocco (1973), who show how a unique generation time distribution function follows from it. This work is related to ensemble methods in the study of populations, which is reviewed elsewhere [Goel, Maitra, and Montroll (1971); see also Sacher (1956), Weiss (1968)].

A stochastic model similar in spirit to the above approach is that of Bronk *et al.* (1968). Their model, which is especially designed for numerical solution with the aid of a computer, divides the cell cycle into a finite number of age intervals. These properties also characterize the "state vector" model of Hahn (1966, 1970), and the models of Takahashi (1966, 1968) and Shackney (1973).

A concurrent theoretical development has been the use of "renewal theory" to represent the total population $N(t)$. Thus, the total population $N(t)$ can be expressed in the present notation and in the absence of cell deaths by the renewal equation [Harris,

1959]

$$N(t) = N_0 \int_t^{\infty} g(\tau)\, d\tau + 2\int_0^t N(t-\tau)\, g(\tau)\, d\tau, \tag{43}$$

where $\int_t^{\infty} g(\tau)\, d\tau$ represents the probability that a cell has a lifetime $\tau > t$. Note that the corresponding expression of the renewal concept in SRF theory is the boundary condition.

From (43) we obtain by differentiation a related integral equation for $dN(t)/dt = \dot{N}(t)$ that has been proposed by Hirsch and Engelberg (1965), namely,

$$\dot{N}(t) = 2\int_{-\infty}^{t} g(t-\tau)\dot{N}(\tau)\, d\tau. \tag{44}$$

Here $g(t-\tau)\, d\tau$ is the probability that a cell of age $t-\tau$ will divide in the time interval t to $t + dt$. The equivalence of (43) and (44) requires that $N(\tau) = N_0\, \delta(\tau)$ for $\tau \leqslant 0$. Otherwise [Martinez, 1966], assuming the number of cells at time $t = 0$ of age $\leqslant -\tau$ is given for $\tau \leqslant 0$ by some distribution function, a generalization of equation (43) is needed to show equivalence. Martinez (1966) has also shown that equation (43) or its generalization, and therefore (44) too, can be derived from the SRF equation, with $\lambda_d = 0$. Thus, he writes

$$N(t) = \int_{-\infty}^{0} E(\tau, t)\, d\tau + 2\int_0^t A(\tau, t)\, \frac{dN(\tau)}{d\tau}\, d\tau, \tag{45}$$

where $E(\tau, t)\, d\tau$ is the expected number of cells at time t that were born in the time interval $\tau - d\tau$ to τ. The function E is expressed in terms of the cell density function n of the SRF equation as

$$E(\tau, t) = n(t-\tau, t). \tag{46}$$

In equation (45), $A(\tau, t)$ is the probability that a cell born in the time interval $\tau - d\tau$ to τ will not divide before time t, and is given as

$$A(\tau, t) = \exp\left[-\int_{\tau}^{t} \lambda_m(\xi - \tau, \xi)\, d\xi\right]. \tag{47}$$

The application of the renewal concept underlies the work of Lotka (1907, 1913, 1922, 1925, 1931, 1945; Sharpe and Lotka, 1911). Present day investigators of demography theory have been made aware of the applicability of the SRF equation to human populations [Fredrickson (1971); Keyfitz (1972)] although it does not appear that the equation has been exploited in that context to any significant extent.

7. MATURATION-STRUCTURED THEORIES

The idea of representing various physiological properties of cells such as size or mass, rather than age, as independent variables in a mathematical theory of cell populations has occurred to numerous investigators. In fact, observations of distribution curves for cell size motivated Scherbaum and Rasch (1957) to introduce their equation. Walker (1954) had already proposed, before the advent of the above work, that the amount of DNA in a cell could be utilized as a measure of the cell's relative position in the cell cycle. Observations of lengths of microorganisms grown in continuous culture led Maclean and Munson (1961) to derive the steady state length distribution of an idealized population in which all cells have an initial length 1 and a final length 21. The growth rate of cells of a given size and its relation to the steady-state length distribution function of all cells, of cells that are newly born, and of cells that are about to divide, was given on the basis of mass balance by Collins and Richmond (1962). Their equation has been rederived subsequently in various ways [Harvey, Marr and Painter (1967); Bell and Anderson (1967); Ramkrishna, Fredrickson, and Tsuchiya (1968)]. Koch and Schaechter (1962) suggested that cell mass, which is essentially equivalent to cell volume, be utilized as an independent variable. They derived the steady-state mass distribution of a growing cell population. This work was refined and extended by investigators of microbial growth [Powell (1964); Koch (1966); Harvey, Marr, and Painter (1967); Kubitschek (1969)]. See Painter and Marr (1968) for a review of mathematical work relating especially to microbial populations.

Subsequently, partial differential equations of the SRF type which utilized a "maturation" variable, were introduced by a number of investigators. Thus Stuart and Merkle (1965) [see also Stuart (1966)] suggested a cell density function $n(\mu, t)$, dependent on "physiological age" μ and time t, that satisfied the equation

$$\frac{\partial n}{\partial t} + v_0 \frac{\partial n}{\partial \mu} = D \frac{\partial^2 n}{\partial \mu^2}. \tag{48}$$

Here v_0 is a mean velocity of maturation and D is a "diffusion constant," which characterize a given population. The diffusion term introduces dispersion into the velocities of maturation of members of the population. An equation similar to (48) was proposed by Sacher (1956) as the basis of a theory of aging.

Oldfield (1966) suggested a continuity equation for a cell density function that depends on any number of maturation variables. A similar proposal that invokes size (mass or volume) and time as variables has been developed by others [Eakman, Fredrickson, and Tsuchiya (1966); Fredrickson, Ramkrishna, and Tsuchiya, 1967]. The equation of Collins and Richmond (1962) has been derived from the latter theory [Ramkrishna, Fredrickson, and Tsuchiya (1968)]. In a simple form of these theories that has been proposed [Rubinow (1968)], the cell density function $n(\mu, t)$ depends on the time t and a maturation variable μ whose biological meaning is left unspecified. The density function n is assumed to satisfy the equation

$$\frac{\partial n}{\partial t} + \frac{\partial}{\partial \mu}(vn) = -\lambda_d n, \quad \mu_0 < \mu < \mu_1, t > 0, \tag{49}$$

where v is the velocity of maturation, a function of μ (and perhaps t) that needs to be specified. Here μ_0 is the initial level of maturation of new born cells (which could for convenience be taken to be zero), and μ_1 is the final level of maturation, when cells divide. Note that the loss function λ_d accounts for cell deaths only and not loss due to cell division. Thus setting v = constant does not reduce equation (49) to the SRF equation. In addition, the boundary and initial conditions for the two equations differ. Thus, for equation (49), the initial condition reads

$$n(\mu, 0) = \varphi(\mu), \tag{50}$$

where $\varphi(\mu)$ is a given function of μ, while the boundary conditon reads

$$v(\mu_0)n(\mu_0, t) = 2v(\mu_1)n(\mu_1, t). \tag{51}$$

The solution of equation (49) by means of the generation expansion has been given. When $\mu_0 = 0$, $\mu_1 = 1$, $v = 1/\tau$, and $\mu\tau = a$, equation (49) becomes identical with the equation for $n(a, t; \tau)$ discussed in section 6, provided that $K(\tau, \tau')$ in equation (41) is set equal to the Dirac delta function $\delta(\tau - \tau')$.

Let μ be thought of as representing size, and assume the idealized case for which $\mu_1 = 2\mu_0$. Assume further that v depends in a linear fashion on size, so that $v = v_0\mu$, where v_0 is a constant. Then the solution of equations (49) and (51) representing a steady state of exponential growth is readily found to be [Bell and Anderson (1967)]

$$n = n_0 e^{\gamma t}\left(\frac{\mu_0}{\mu}\right)^2, \tag{52}$$

where n_0 is a constant. The inverse square dependence of n on size is that given by Maclean and Munson (1961). [See also Koch and Schaechter (1962), Painter and Marr (1968).]

It has also been suggested that the cell density function depends on age a, time t, and a volume variable μ. Thus, Bell [Bell and Anderson (1967); Bell (1968)] proposed a cell density function $n = n(a, \mu, t)$ that satisfies the equation

$$\frac{\partial n}{\partial t} + \frac{\partial n}{\partial a} + \frac{\partial}{\partial \mu}(vn) = -(\lambda_m + \lambda_d)n. \tag{53}$$

Here the number of cells in the age interval a to $a + da$ and volume interval μ to $\mu + d\mu$ at time t is given by $n(a, \mu, t)\, da\, d\mu$. The loss functions due to cell division and cell death, λ_m and λ_d

respectively, and the velocity function v, are assumed to depend on a and μ only. The birth condition is expressed as

$$n(0, \mu, t) = 4\int_0^\infty \lambda_m(\tau, 2\mu)n(\tau, 2\mu, t)\, d\tau. \tag{54}$$

The factor of 4 on the right-hand side occurs because of two factors of 2: one for cell division and one because cells leave a volume interval $2d\mu$ and enter a volume element $d\mu$. A unique solution of equation (53) requires also the specification of the initial age-volume distribution of the population $n(a, \mu, 0)$. The solution to equation (53), the derivation of the Collins-Richmond equation from it, the steady-state exponential growth behavior, and the application of it to a number of experimental observations are all discussed in the above-mentioned references and some additional ones [Anderson and Petersen (1967); Anderson *et al.* (1969); Anderson and Bell (1971)].

Weiss (1968) proposed an integro-differential equation for $n = n(a, \mu, t)$ that is Markoffian in spirit. In the Fokker-Planck approximation, in which the change in the physiological age (maturation variable) is assumed to be small in the time interval t to $t + dt$, the equation simplifies to a partial differential equation, which reduces in a special case to that of Stuart and Merkle (1965).

ADDENDUM

Note added in proof: Because the renewal principle was known already in the early part of this century, it should not surprise us that an earlier date for the discovery of the SRF equation exists. In fact such is the case: I am indebted to Frank Hoppensteadt for pointing out to me that equation (7) already appeared in 1926 in the work of A. G. M'Kendrick [*Proc. Edinburgh Math. Soc.* **44**, 98]. M'Kendrick's interest was in the theory of epidemics. This circumstance suggests that the SRF equation should properly be called the M'Kendrick equation.

REFERENCES

1. E. C. Anderson and D. F. Petersen, *Biophys. J.*, **7** (1967), 353.
2. E. C. Anderson, G. I. Bell, D. F. Petersen, and R. A. Tobey, *Biophys. J.*, **9** (1969), 246.
3. E. C. Anderson and G. I. Bell, *Biophys. J.*, **11** (1971), 308.
4. J. C. Barrett, *J. Nat. Cancer Inst.*, **37** (1966), 443.
5. G. I. Bell, *Biophys. J.*, **8** (1968), 431.
6. G. I. Bell and E. C. Anderson, *Biophys. J.*, **7** (1967), 329.
7. B. V. Bronk, G. J. Dienes, and A. Paskin, *Biophys. J.*, **8** (1968), 1353.
8. J. F. Collins and M. H. Richmond, *J. Gen. Microbiol.*, **28** (1962), 15.
9. A. G. De Rocco and W. H. Woolley, *Math. Biosc.*, **17** (1973), 666.
10. J. M. Eakman, A. G. Fredrickson, and H. M. Tsuchiya, *Chem. Eng. Prog. Symp. Series No. 69*, **62** (1966), 37.
11. J. Engelberg, *Exp. Cell Res.*, **36** (1964), 647.
12. A. G. Fredrickson, *Math. Biosc.*, **10** (1971), 117.
13. A. G. Fredrickson, D. Ramkrishna, and H. M. Tsuchiya, *Math. Biosc.*, **1** (1967), 327.
14. N. S. Goel, S. C. Maitra, and E. W. Montroll, *Rev. Mod. Phys.*, **43** (1971), 231.
15. G. M. Hahn, *Biophys. J.*, **6** (1966), 275.
16. ———, *Math. Biosci.*, **6** (1970), 295.
17. T. E. Harris, *The Kinetics of Cellular Proliferation*, F. Stohlman, Jr., ed., Grune & Stratton, New York, 1959, 368.
18. J. D. Harvey, *J. Gen. Microbiol.*, **70** (1972), 109.
19. R. J. Harvey, A. G. Marr, and P. R. Painter, *J. Bacteriol.*, **93** (1967), 605.
20. H. R. Hirsch and J. Engelberg, *J. Theor. Biol.*, **9** (1965), 297.
21. W. H. Hughes, *J. Gen. Microbiol.*, **12** (1955), 265.
22. C. D. Kelly and O. Rahn, *J. Bacteriol.*, **23** (1932), 145.
23. D. G. Kendall, *Biometrika*, **35** (1948), 316.
24. N. Keyfitz, *Proc. Sixth Berkeley Symposium on Mathematical Statistics and Probability*, vol. 4, L. M. LeCam, J. Neyman, and E. L. Scott, eds., University of California Press, Berkeley, 1972, 89.
25. A. L. Koch, *J. Gen. Microbiol.*, **45** (1966), 409.
26. A. L. Koch and M. Schaechter, *J. Gen. Microbiol.*, **29** (1962), 435.
27. H. E. Kubitschek, *Exp. Cell Res.*, **26** (1962), 439.
28. ———, *Nature*, **209** (1966a), 1039.

29. ———, *Exp. Cell Res.*, **43** (1966b), 30.

30. ———, *Proc. Fifth Berkeley Symposium on Mathematical Statistics and Probability*, vol. 4, L. M. LeCam and J. Neyman, eds., University of California Press, Berkeley, 1967, 549.

31. ———, *Biophys. J.*, **9** (1969), 792.

32. ———, *Cell Tissue Kinet.*, **4** (1971), 113.

33. J. L. Lebowitz and S. I. Rubinow, *J. Math. Biol.*, **1** (1974), 17.

34. A. J. Lotka, *Amer. J. Sci.*, **24** (1907), 199.

35. ———, *J. Wash. Acad. Sci.*, **3** (1913), 241, 289.

36. ———, *Proc. Nat. Acad. Sci.*, **8** (1922), 339.

37. ———, *Elements of Physical Biology*, Williams and Wilkins, Baltimore, 1925. (Republished as *Elements of Mathematical Biology*, Dover, New York, 1956.)

38. ———, *Human Biol.*, **3** (1931), 459.

39. ———, *Essays on Growth and Form Presented to D'Arcy Wentworth Thompson*, W. E. L. Clark and P. P. Medawer, eds., Oxford University Press, Oxford, 1945, 355.

40. F. I. Maclean and R. J. Munson, *J. Gen. Microbiol.*, **25** (1961), 17.

41. H. M. Martinez, *Bull. Math. Biophys.*, **28** (1966), 411.

42. D. S. Nachtwey and I. L. Cameron, *Cell Cycle Analysis*, in *Methods in Cell Physiology*, D. M Prescott, ed., Academic Press, New York, 1968, 214.

43. G. C. Nooney, *Biophys. J.*, **7** (1967), 69.

44. D. G. Oldfield, *Bull. Math. Biophys.*, **28** (1966), 345.

45. P. R. Painter and A. G. Marr, *J. Gen. Microbiol.*, **48** (1967), 155.

46. ———, *Ann. Rev. Microbiol.*, **22** (1968), 519.

47. E. O. Powell, *Biometrika*, **42** (1955), 16.

48. ———, *J. Gen. Microbiol.*, **37** (1956), 281.

49. ———, *J. Gen. Microbiol.*, **18** (1958), 382.

50. ———, *J. Gen. Microbiol.*, **37** (1964), 231.

51. ———, *J. Gen. Microbiol.*, **58** (1969), 141.

52. E. O. Powell and F. P. Errington, *J. Gen. Microbiol.*, **31** (1963), 315.

53. D. M. Prescott, *Exp. Cell Res.*, **10** (1959), 279.

54. O. Rahn, *J. Gen. Physiol.*, **15** (1932), 257.

55. D. Ramkrishna, A. G. Fredrickson, and H. M. Tsuchiya, *Bull. Math. Biophys.*, **30** (1968), 319.

56. S. I. Rubinow, *Biophys., J.*, **8** (1968), 1055.

57. G. Sacher, *Radiology*, **67** (1956), 250.

58. M. Schaechter, J. P. Williamson, J. R. Hood, Jr., and A. L. Koch, *J. Gen. Microbiol.*, **29** (1962), 421.

59. O. Scherbaum and G. Rasch, *Acta Pathol. Microbiol. Scand.*, **41** (1957), 161.

60. S. E. Shackney, *J. Theor. Biol.*, **38** (1973), 305.

61. F. R. Sharpe and A. J. Lotka, *Phil. Mag.*, **21** (1911), 435.

62. R. N. Stuart, *The Kinetics of Cell Growth*, UCRL-70039, University of California Radiation Laboratory, Livermore, 1966.

63. R. N. Stuart and T. C. Merkle, *Calculation of Treatment Schedules for Cancer Chemotherapy, Part II*, UCRL-14505, University of California Radiation Laboratory, Livermore, 1965.

64. M. Takahashi, *J. Theor. Biol.*, **13** (1966), 202.

65. ———, *J. Theor. Biol.*, **18** (1968), 195.

66. E. Trucco, *Bull. Math. Biophys.*, **27** (1965), 449.

67. ———, *Bull. Math. Biophys.*, **32** (1970), 459.

68. P. F. Verhulst, *Mem. Acad. Roy. Bruxelles*, **18** (1845), 1.

69. ———, *Mem. Acad. Roy. Bruxelles*, **20** (1847), 1.

70. H. von Foerster, *The Kinetics of Cellular Proliferation*, F. Stohlman, Jr., ed., Grune & Stratton, New York, 1959, 382.

71. P. M. B. Walker, *J. Exp. Biol.*, **31** (1954), 8.

72. G. Weiss, *Bull. Math. Biophys.*, **30** (1968), 427.

73. E. O. Wilson and W. H. Bossert, *A Primer of Population Biology*, Sinauer Associates, Sunderland, Mass., 1971.

74. W. H. Woolley and A. G. De Rocco, *J. Theor. Biol.*, **39** (1973), 73.

THE DYNAMICS OF NONLINEAR MODELS WITH AGE STRUCTURE*

G. Oster

INTRODUCTION

There are three equivalent models commonly used to describe the age structure of a population: (e.g., Oster, 1976):

1. The so-called von Foerster equation:

$$\frac{\partial n}{\partial t} + \frac{\partial n}{\partial a} = -\mu n \tag{1a}$$

$$n(t, 0) = \int_{\alpha}^{\alpha+\gamma} mnda \tag{1b}$$

where $n(t, a)$ is the density function giving the number of individuals at time t and age a. $\mu(\cdot)$ is the (per capita) deathrate, and $m(\cdot)$ the (per capita) birthrate; both may be functions of t, a. α is the earliest age where reproduction commences and γ is the width of the "reproductive window".

2. If equation (1a) is solved formally for $n(t, a)$ and substituted in (1b) the resulting integral equation for the total birth-

*Research supported by NSF Grant no. BMS 74-21240.

rate $B(t) = n(t, 0)$ is called Lotka's equation:

$$B(t) = \int_{\alpha}^{\alpha+\gamma} m(a) l(a) B(t-a) da \tag{2}$$

where $l(a) = \exp[-\int_0^a \mu da']$ is the fraction of individuals surviving to age a.

3. If we finite difference equations (1) forward in time and backward in age the resulting difference equations can be written as:

$$n_{t+1,a} = (1 - \mu(t, a-1))n_{t,a} \equiv S_{t,a-1} n_{t,a}, a = 0, 1, \ldots, A, \tag{3a}$$

$$n_{t,0} = \sum_{a=0}^{A} \overline{m}_{t-b,a} n_{t-1,a}, \tag{3b}$$

where $S_{t,a-1}$ is the fraction of individuals surviving from age class $a-1$ to age class a and $\overline{m}_a$ is the average fecundity of individuals in age class a. Equations (3), written in matrix form, are called the Leslie model:

$$\mathbf{n}_{t+1} = \mathbf{L}\mathbf{n}_t \tag{4}$$

where

$$\mathbf{n}_t = (n_{t,0}, n_{t,1}, \ldots, n_{t,A})^T, \tag{5a}$$

$$\mathbf{L} = \begin{bmatrix} m_0 & m_1 & \cdots & m_A \\ S_1 & 0 & & \\ & S_2 & 0 & \\ 0 & & S_{A-1} & 0 \end{bmatrix}. \tag{5b}$$

When the birth and death rates are functions of age and time only then the models are linear and solutions can be written explicity whose properties have been well studied. In this chapter we shall focus on some features of the age equations when nonlinear terms are included. Gurtin and MacCamy (1974) and

Swick (1976) have studied the existence and uniqueness of solutions to equations (1) when the birth and death rates are functionals of the total population, $N(t) = \int_0^\infty n da$. Here, we shall use the discrete model (4), but many of the results will generalize to the infinite dimensional case.

2. BIFURCATION BEHAVIOR OF SINGLE-POPULATION MODELS

2.1. Figure 1 shows some data collected by the Australian entomologist A. J. Nicholson on the population dynamics of the sheep blowfly. Nicholson maintained a population of flies on a constant, but restricted protein diet. The competition for food when the population levels were high meant that most flies did not receive enough protein to realize their maximum fecundity. Thus few eggs were laid and the population crashed. The subsequent generation, being smaller, faced less intense competition for food and their fecundity rebounded. The resulting alternation of large and small generations produced the oscillations with a period of about 38 days evident in Figure 1.

A closer examination of the data reveals, in addition to the major cycle, several other features which beg explanation: (i) there is a "high frequency" component superimposed on the major cycles causing a "double peak", (ii) a "low-frequency" envelope which spans 3 of the major peaks, (iv) after the 400th day the

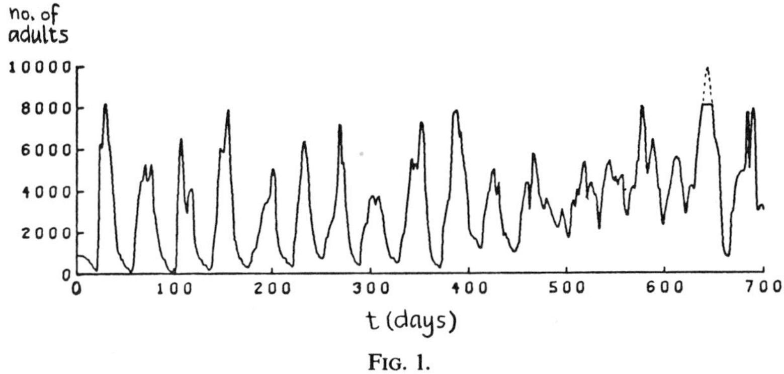

FIG. 1.

pattern of peaks becomes much more chaotic in appearance, although still retaining strong traces of the oscillatory behavior. In this section we shall show how the qualitative aspects of all of these features of the data show up in the dynamics of rather simple population models. A more detailed study can be found in Oster, Auslander and Allen (1975), where more complicated models are used which can fit the data quantitatively as well as qualitatively.

2.2. There are two mechanisms which underlie the dynamics exhibited by Nicholson's flies: (1) the maturation delay from egg to reproducing adult and (2) the nonlinear dependence of the adult fecundity on the limited food supply. We can model both of these effects using the von Foerster model (1a, b), if we assume that because of competition for food the birthrate $b(\cdot)$ is a decreasing function of the current adult population; for example:

$$b(a, N) = (b_0 - b_1 N) H(a - \alpha) \tag{6a}$$

or

$$= b_0 e^{-b_1 N} H(a - \alpha) \tag{6b}$$

or

$$= \frac{b_0}{1 + b_1 N} H(a - \alpha), \tag{6c}$$

where $N(t) = \int_\alpha^\infty n\,da$ is the total number of adults and $H(a - \alpha)$ is a step function at $a = \alpha$, the earliest age of reproduction. Using such a form for the birthrate in equation (2) and integrating (1) over all adult ages yields a functional equation in $N(t)$ (cf. Oster, 1976a, b);

$$\begin{aligned} \frac{dN}{dt} &= e^{-\mu\alpha} b[N(t - \alpha)] N(t - \alpha) - \bar{\mu} N(t) \\ &\triangleq B[N(t - \alpha)] - \bar{\mu} N(t). \end{aligned} \tag{7}$$

The birthrate $B[\cdot]$ is a nonlinear function of the population density-delayed by the maturation time, α. An analysis of equation

(7) shows that it exhibits a stable limit cycle with period about 2α. Moreover, because the birthrate function $B[\cdot] = b(N)N$ has a critical point every oscillation in the adult population, providing its amplitude exceeds the critical point in $b[N]N$, will produce a periodic burst of egg laying whose waveform has two peaks. If the preadult mortality is sufficient the ensuing adult peak will also have the double peak observed in the data of figure 1. Rather than pursuing our analysis using this model, however, we shall proceed by a simpler approximation. If we discretize equation (1a, b) using one of the birthrate functions (6) we obtain a nonlinear Leslie model:

$$\mathbf{n}_{t+1} = \mathbf{L}(N)\mathbf{n}_t,$$
$$N_t = \mathbf{1}^T\mathbf{n}_t = \Sigma n_{i_t}. \tag{8}$$

To simplify matters still further we consider a 2–age class model:

$$\begin{pmatrix} n_1 \\ n_2 \end{pmatrix}_{t+1} = \begin{pmatrix} b_1(N) & b_2(N) \\ s & 0 \end{pmatrix}\begin{pmatrix} n_1 \\ n_2 \end{pmatrix}_t,$$
$$N_t = n_{1_t} + n_{2_t}. \tag{9}$$

2.3. The dynamics of equation (9) are extremely complex and we shall give a qualitative description of its behavior in the next section. Before proceeding, however, it is enlightening to study equation (9) in the special case when $b_1 = 0$. Then, notice that if we examine equation (9) at every other time step $(0, t + 2, t + 4, \ldots,)$ we have

$$\begin{pmatrix} n_1 \\ n_2 \end{pmatrix}_{t+1} = \mathbf{L}(N)^2\begin{pmatrix} n_1 \\ n_2 \end{pmatrix}_t$$
$$= \begin{pmatrix} sb_1 & 0 \\ 0 & sb_2 \end{pmatrix}\begin{pmatrix} n_1 \\ n_2 \end{pmatrix}_t. \tag{10}$$

That is, equation (10) is equivalent to the single nonlinear difference equation

$$X_{t+1} = f(x_t) \\ = sb[n_1 + n_2]X_t, \tag{11}$$

where X_t is either n_{1_t} or n_{2_t} and now the time step for (11) is taken as two time steps for (10). Thus we are led to examine nonlinear difference equations of the general shape shown in Figure 2a:

$$f:\mathbf{R}^1 \to \mathbf{R}^1, \\ x \mapsto f(x), \tag{12}$$

where f has a single critical point $f^1(x_c) = 0$. For concreteness, we shall discuss briefly the dynamics of equation (12) when the birthrate nonlinearity is quadratic (i.e., using 6a), so that we have a difference equation of the form (cf. Fig. 2b):

$$X_{t+1} = rX_t(1 - x_t) \equiv f_r(x_t). \tag{13a}$$

2.4. When the net reproductive rate parameter, r, is less than 1 the origin is the only fixed point for $f(\cdot)$. For $1 < r < 2$ there is a nonzero fixed point at $\bar{x} = 1/r$ to which the orbit is attracted monotonically. When $r > 2$, $\bar{x} > x_c$ and the orbit decays to $\bar{X}$ in an oscillatory fashion. The most interesting dynamic behavior occurs when $3 \leqslant r \leqslant 4$.

The local stability of the equilibrium $\bar{X}$ is governed by the eigenvalue (slope) of $f(\cdot)$ at $\bar{X}$:$f'\bar{x}) = 2 - r$. When $r < 3$, $\bar{x}$ is attracting, but at $r = 3$, $f'(\bar{x}) = -1$ and $\bar{x} = 2/3$ becomes unstable and the orbit winds out from $\bar{x}$.

To see where the orbit goes we must examine the second iterate of $f(\cdot)$, i.e., $f \circ f(x) \equiv f^2(x) = 9x - 36x^2 + 54x^3 - 27x^4$. The graph of f^2 is obtained by plotting x_t vs. x_{t+2} as shown in Figure 3. Since $f^{2'}(2/3) = 1$, $f^{2''}(2/3) = 0$, and $f^{2'''}(2/3) < 0$, the graph of f^2 crosses the 45 degree tangentially from above as shown in Figure 3a. As r increases past 3, f^2 intersects the 45 degree line at 3 points as shown in Figure 3b. The original fixed point at $\bar{x}$ has bifurcated into two points which form a period-2 orbit as shown in figure 3c.

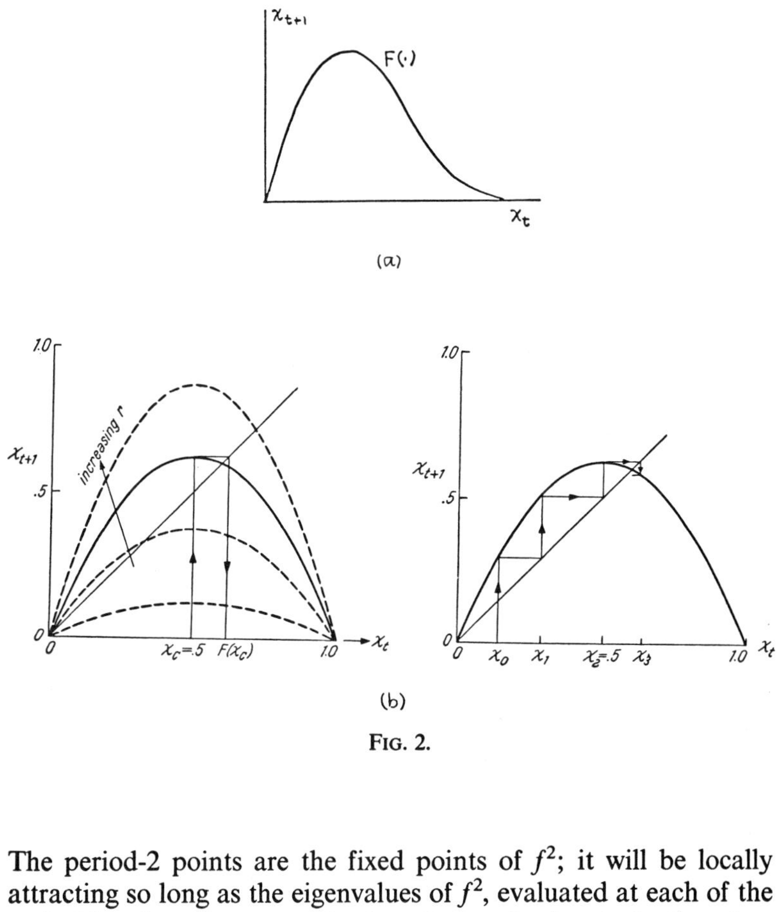

FIG. 2.

The period-2 points are the fixed points of f^2; it will be locally attracting so long as the eigenvalues of f^2, evaluated at each of the period-2 points, have modulus less than unity (equivalently, when the product of the slopes of f^2 at the two fixed points has absolute value less than unity).

As r is increased the period-2 orbit eventually becomes unstable, splitting into a period-4 orbit. Cycles of length 2^k, $k = 0, 1, 2 \ldots,$ are generated this way as the parameter r increases; however, the parameter intervals corresponding to successively higher periods become progressively smaller. Eventually, a critical value, r^*, is reached whereupon a bifurcation of a wholly different nature occurs. For values of r above r^* many periodic trajectories coexist,

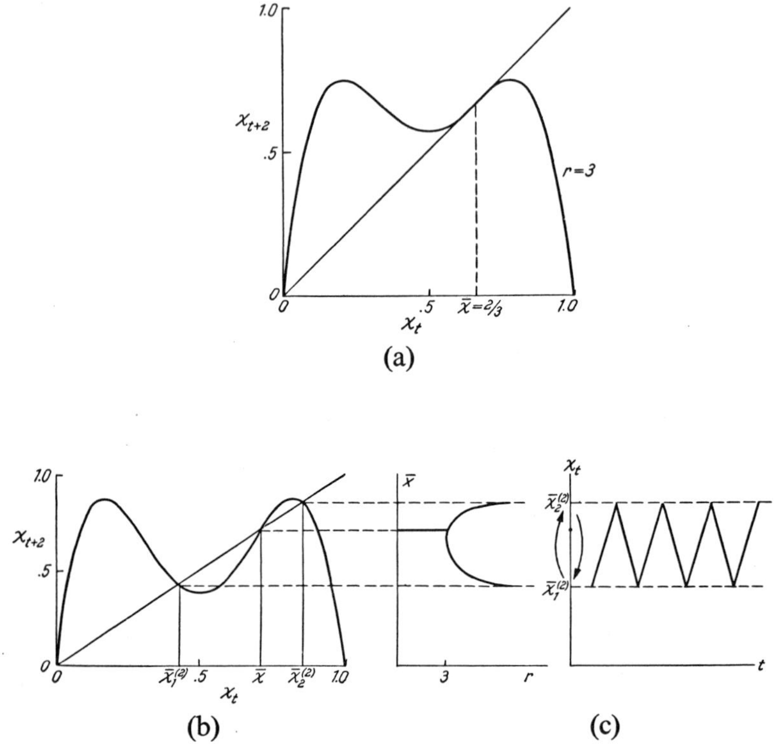

FIG. 3.

although it can be shown that only one of these periodic orbits is stable at a time (Guckenheimer, Oster, and Ipaktchi, 1976). Fixing the precise value of r^* is difficult, although Li and Yorke (1976) proved that by the time $r = r_3$ where a period-3 orbit is observed, the orbits of arbitrary length are present. It is possible to devise a scheme for counting the number of coexisting orbits of a particular period for a given value of the parameter (Smale and Williams, 1976). For the parameter region near the period-3 orbit it turns out that N_k, the number of fixed points of period k, follows a Fibonacci sequence: $N_k = \{1, 3, 4, 7, 11, 18, \ldots\}$;e.g., there are 18 fixed points of f^6, and so $(18 - 1 - 2 - 3)/6 = 2$ periodic orbits of least period 6. The specification of which values of r lead to which stable periodic orbits is poorly understood, although for low periods some of the corresponding parameter values are calculated in Metropolis Stein and Stein, 1973.

2.5. Even more interesting than the periodic orbits is the fact that, by the time the period-3 orbit makes its appearance, along with the cycles of arbitrary length there appear orbits which are completely aperiodic. Fortunately, in most 1-dimensional models these chaotic-appearing orbits occur only for isolated initial conditions in the interval [0, 1], so they are—in some sense—rare. As we shall see this is not the case in higher dimensions. For all practical purposes, however, it is not possible to distinguish an aperiodic orbit from a segment of a long periodic orbit. The significance of chaotic-appearing orbits for interpreting real data in terms of these sorts of models is apparent (May and Oster, 1976).

It is certainly surprising that so simple a deterministic model can exhibit such complex dynamic behavior. To see just how bad things can get we can examine the extreme case when $r = 4$. For this choice of r it turns out that the entire interval [0, 1] generates chaotic orbits, and so the only way to characterize the dynamic behavior is statistically. One way of doing this is the following. The function $f_4(x) = 4x(1 - x)$ is topologically equivalent to the triangle function $\hat{f}(x)$ shown in Figure 4 using the coordinate change $h(x) = (2/2\pi) \sin^{-1}\sqrt{x}$ (that is, orbits of f_4 are homeomorphic to orbits of $\hat{f}$ under $h(\cdot)$). If we agree to count as "0" points on the orbit of $\hat{f}$ that land in the left half of the unit interval and count as "1" orbits which land in the right half of the interval, then $\hat{f}^k$ generates a sequence of 0's and 1's of length k. It can be shown that this binary sequence is indistinguishable from a series of Bernoulli trials, (Guckenheimer, Oster, and Ipaktchi, 1976). Moreover, the orbits of $\hat{f}$ distribute their points uniformlly over the unit interval. If we map this uniform density back with $h^{-1}(\cdot)$ we obtain a density function, $\rho = (1/\pi)/\sqrt{x(1 - x)}$ which can be used to compute the asymptotic probability of finding the orbit in any subinterval in [0, 1]. This technique of describing the dynamics of f statistically can be extended to the higher dimensional case, to which we now turn.

2.6. As one might expect, the 2-age-class model equation (9) displays an even more complex spectrum of dynamics than the

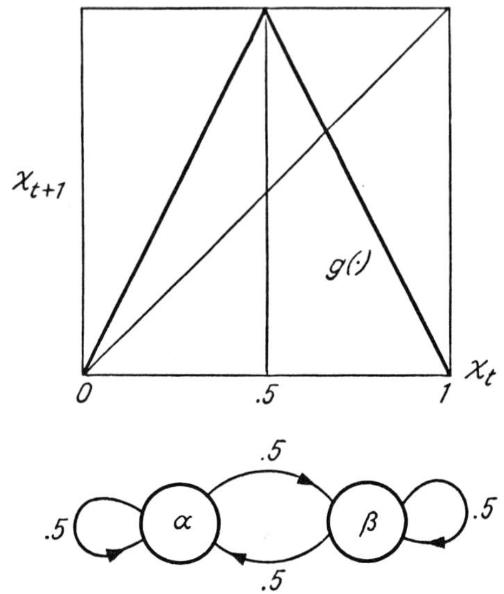

FIG. 4.

single age class version. For purposes of exposition let us specialize equation (9) by choosing $b_1[N] = b_2[N] = b_0 e^{-b_1 N}$ and $S = 1$. Thus the model becomes:

$$\begin{aligned} &\mathbf{F}:\mathbf{R}^2 \to \mathbf{R}^2 \\ &(x, y) \mapsto \left[b_0(x+y)e^{-b_1(x+y)}, X_1\right]. \end{aligned} \tag{13b}$$

If we fix b_1, say $b_1 = 0.1$, we can study the dynamics of $\mathbf{F}$ as the parameter b_0 is varied.

Figure 5 shows a sequence of phase plane and time plots for various values of b_0. For $b_0 \lesssim 8.95$ there is a unique, globally attracting equilibrium at $\bar{x} = \bar{y} = \ln(2b_0/2b_1)$. At $b_0 \approx 8.95$ a 3-point cycle appears at finite amplitude surrounding the equilibrium. An examination of the characteristic equation of equation (13), after linearizing about $(\bar{x}, \bar{y})$, reveals that the equilibrium point remains locally attracting until $b_0 = e^3/2 \approx 10.05$. Thus, for $8.95 < b < e^3/2$ the stable 3-point cycle coexists with a stable equilbrium point. The bifurcation which produced

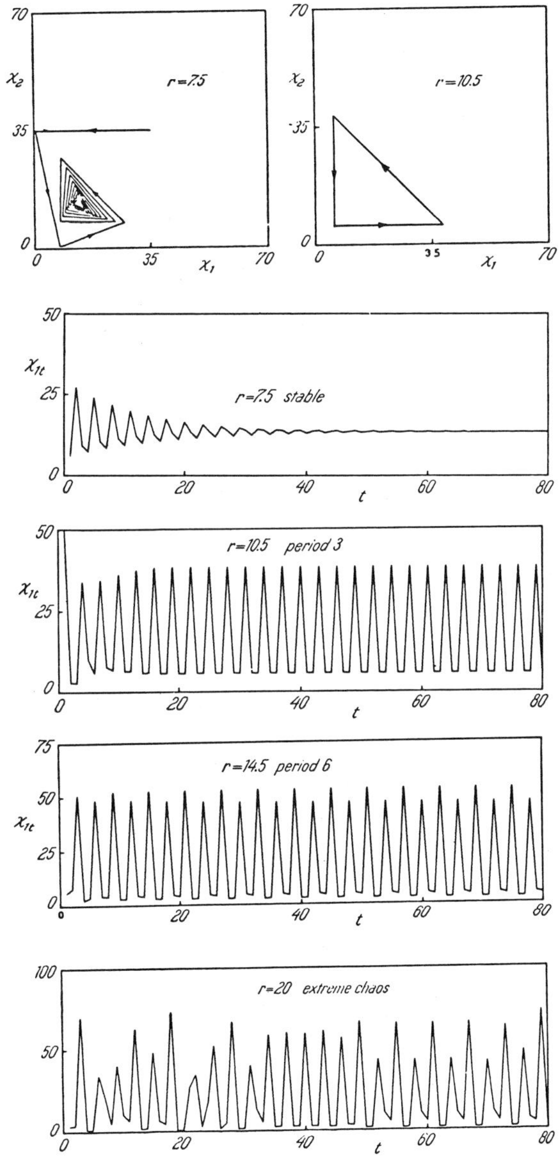

r=7.5
r=10.5
70
35
0
x_2
x_1
x_{1t}
r=7.5 stable
r=10.5 period 3
r=14.5 period 6
r=20 extreme chaos
t
0
20
40
60
80
25
50
75
100

FIG. 5.

this 3-point cycle is of a different nature than the period-doubling bifurcations in the single age-class model. There the bifurcation occurred when the eigenvalue of the characteristic equation passed out of the unit circle $|z| = 1$ along the real axis at $z = -1$. In this case, the eigenvalues of the characteristic equation are complex and pass through the unit circle at the roots of $z^3 = 1$ (i.e., at $2\pi/3$). This accounts for the cycle being of length 3: the Jacobian of $\mathbf{F}$ is a rotation of the plane by $2\pi/3$. The creation of the 3-point cycle is depicted schematically in Figure 6. We see that as b_0 passes through 8.95 a pair of period-3 orbits is created, but the inner one is not stable, and so is unobservable. As $b_0 \to e3/2$ the unstable orbit decreases in amplitude until at $b_0 = e^3/2$ it coalesces with the stable point leaving an unstable equilibrium.

Further increases in b_0 merely increase the amplitude of the 3-point cycle—which is now globally attracting—until $b_0 \approx 14.5$, whereupon a period-doubling bifurcation occurs wherein each of the apex points of the 3-point cycle splits into 2 points, creating a stable 6-point cycle. This doubling occurs by a bifurcation through $z = -1$ exactly as in the 1-dimensional case. As b_0 is increased still further, the 6-point cycle gives way to 12, 24 and 48 . . . point cycles. The parameter intervals corresponding to this sequence of $3 \cdot 2^k$ point cycles decreases, just as in the 1-dimensional model. Near $b_0 \approx 17$ a new kind of bifurcation appears analogous to, but different from, that which occurred at $r = r^*$ in the 1-dimensional case. At this point the orbit of $\mathbf{F}$ commences to appear more and

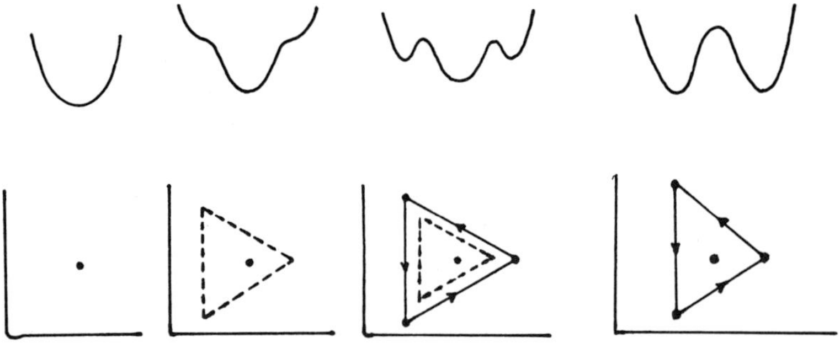

FIG. 6.

more chaotic. This progression of bifurcations is summarized in Figure 7. There we see that if b_0 is increased still further, eventually, near $b_0 \approx 24$ the chaos is replaced by a stable 4-point cycle. Order has reemerged via a bifurcation through $z = +1$ which creates a new fixed point by a process similar to that which created the period-3 orbit in the 1-dimensional case. This 4-point cycle then generates a sequence of $4 \cdot 2^k$ point cycles, via bifurcations through $z = -1$, which eventually terminates in another chaotic regime.

This pattern of alternating bands of chaos and $n \cdot 2^k$-point cycles ($n = 3, 4, 5 \ldots$) appears to continue as shown in Figure 7, with the periodic bands progressively decreasing in width, so that for sufficiently high birthrates chaotic dynamics (or extremely long cycles which are practically indistinguishable) is virtually assured.

There are two features of this dynamic progression which deserve comment. First, in Figure 8 we see that the first chaotic regime coalesces about the apexes of original 3-point cycle. Thus a time series of $\{x_t\}$ or $\{y_t\}$ would appear to be a "noisy" 3-point cycle, i.e., a period-3 oscillation whose amplitude was subject to statistical fluctuations. As b_0 is increased toward the middle of the first chaotic band the density of points in the phase plane spreads so that the periodic component is suppressed. Moving toward the right half of the first chaotic band the density again begins to coalesce about four apexes until the bifurcation through $z = +1$ at $b_0 \approx 23.8$ creates a true 4-point cycle. This pattern, which repeats itself through all the chaotic bands, is a consequence of the somewhat artifical symmetry we have introduced into our model by equating the reproductive rates in both age classes. Nevertheless, it illustrates one type of "macroscopic" pattern that emerges from the complicated dynamics in the chaotic regimes. Another macroscopic feature which becomes pronounced at high birthrates is the pattern of periodic "outbreaks": since each successive bifurcation through $x = +1$ creates a new fixed point near the origin the orbit tends to spend more and more time at low population levels, punctuated by occasional brief excursions and rapid crashes. This same behavior is observed in the 1-age-class model (May, 1975).

If the reproductive rates of the two age classes are not equal

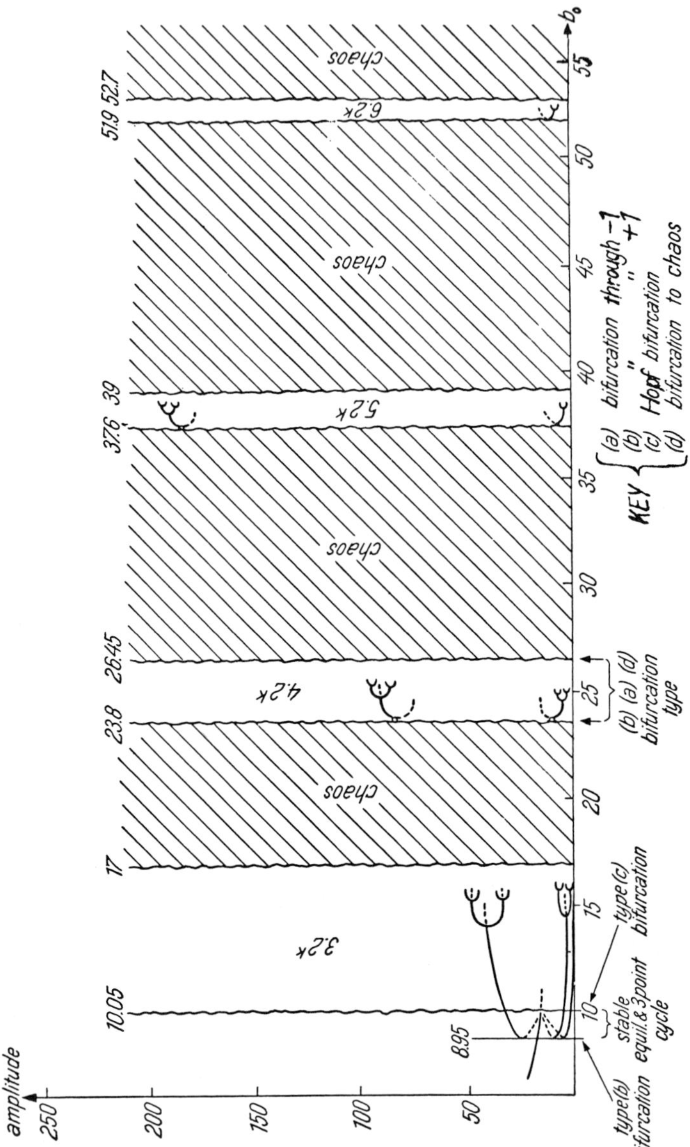
amplitude
250
200
150
100
50
8.95
10.05
17
23.8
26.45
37.6
39
51.9 52.7
10
15
20
25
30
35
40
45
50
55
b_0
3.2k
4.2k
5.2k
6.2k
chaos
type(b) bifurcation
stable equil.& 3point cycle
type(c) bifurcation
(b) (a) (d) bifurcation type
KEY
(a) bifurcation through −1
(b) " " +1
(c) Hopf bifurcation
(d) bifurcation to chaos

FIG. 7.

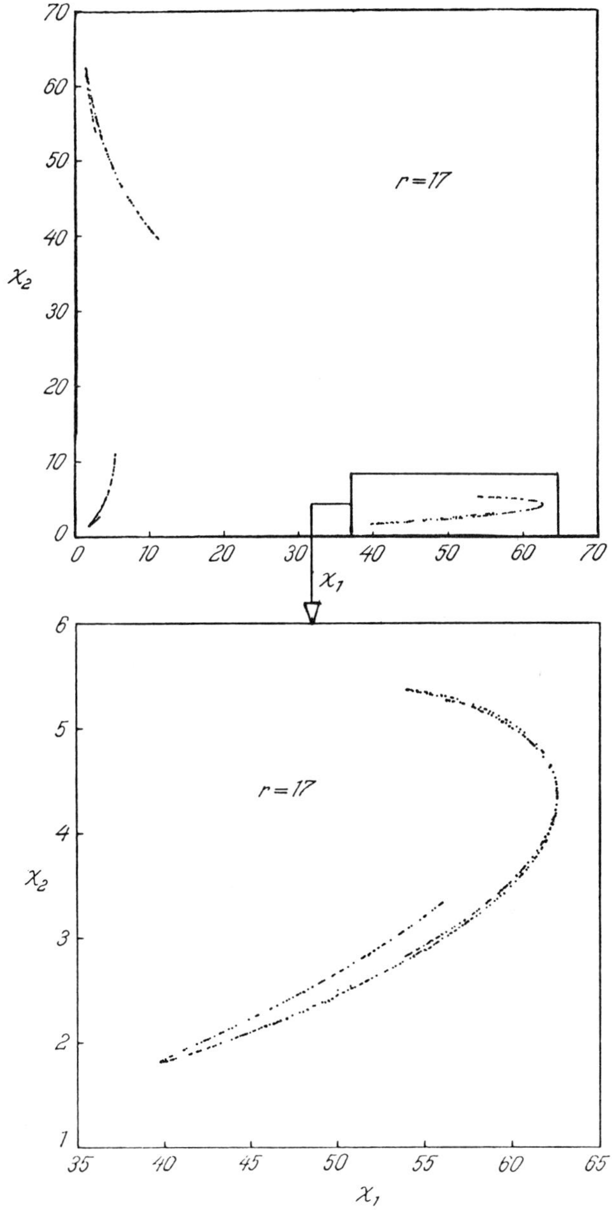

FIG. 8.

then the Jacobian of **F** is not a rotation by a rational angle, i.e., the eigenvalues of $D\mathbf{F}$ pass through $|z| = 1$ at $2\pi p/q$, where q may not be rational. Then the initial bifurcation produces not a periodic orbit, but an "invariant circle" (Guckenheimer, Oster, and Ipaktchi, 1976). That is, the cycle doesn't quite close on itself, but precesses around a closed curve in the phase plane. This is shown in Figure 9 for the case where $b_{01} = 10$ and $b_{02} = 15$. As a consequence of this precession the time series generated appears almost periodic, but with a low frequency "beat" superimposed on the cycle. Figure 10 shows a typical beat pattern generated by the linearized map

$$D\mathbf{F} = \begin{pmatrix} \cos\theta & \sin\theta \\ -\sin\theta & \cos\theta \end{pmatrix}. \tag{14}$$

Of course, the trajectories of the linearized system are neutrally stable (the eigenvalues of (14) are purely imaginary), while the stability of the nonlinear system must be determined by a lengthy calculation from bifurcation theory (Marsden and McCracken, 1976).

The emergence of chaos is more complex when the chaotic orbits nucleate about the invariant circle than when they emerge from the periodic regime, and we shall not discuss this further here.

2.7. To summarize, we see that as the birthrate of the nonlinear population model (13) is increased 4 kinds of bifurcations can occur—each characterized by a distinct set of qualitative dynamical features. (i) Bifurcations wherein an eigenvalue passes through the unit circle at $z = -1$. These double the period of an existing orbit; (ii) bifurcations through $z = +1$ which create a new fixed point. In our model this resulted in the creation of new fixed points near the origin so that the orbit took on the appearance of periodic "outbreak"; (iii) bifurcations through $|z| = 1$ with complex eigenvalues (so called "Hopf" bifurcations). These give rise to either periodic orbits or invariant circles depending on whether the argument of the eigenvalues is rational or irrational. We have seen that, in the former case, one can have periodic orbits coexisting with stable equilibria. In the latter case we saw that the orbits were

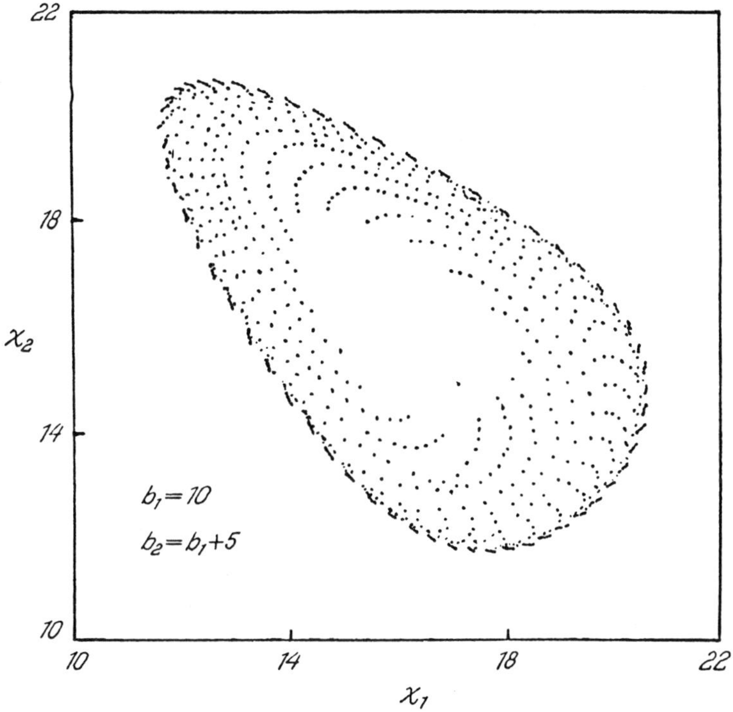

FIG. 9.

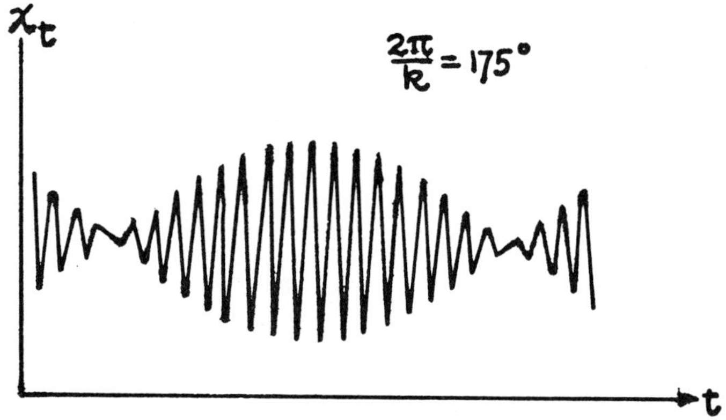

FIG. 10.

characterized by quasi-periodic motion with a low frequency envelope, resembling a "beat"; (iv) finally, there were the bifurcations which gave rise to the chaotic orbits. The nature of these bifurcations are poorly understood at the present time. However, we have seen that the qualitative features of the dynamics in the chaotic regime can be continuously graded between two extremes depending on the value of the bifurcation parameter, the orbit could look like a stable cycle, but with some randomness in the location of the apexes, or it could appear completely random—as if it were a true stochastic process.

Thus, we see that the asymptotic dynamical behavior can be characterized by several different kinds of attracting sets in the phase plane. In addition to the stable equilibria and periodic orbits that are the staple of theoretical ecology there are at least two additional kinds of attracting sets, the invariant circle and the chaotic "strange attractor". Inside the strange attractor the orbits are indeed unpredictable (each point of the attractor is "hyperbolic" i.e., having a stable and an unstable direction). Thus no detailed, deterministic description of the dynamics can be given and so one must resort to a statistical description of the orbits. Such a statistical treatment depends on constructing a probability measure, or orbit density function, on the attractor. This can be done analytically in some cases, or, if that is too difficult, calculated on a computer. Using this density function one can compute the likelihood of finding the orbit in a particular region of phase space (Guckenheimer, Oster, and Ipaktchi, op. cit.).

2.8. Returning now to the blowfly experiments in Figure 1 we can see that the addition to the Leslie model of a density dependent depression in fecundity reproduces, at least qualitatively, the four dynamical features discussed in section 2–1. A careful simulation study, using a more detailed model with many age classes, shows that quantitative agreement between model and data can be achieved (Oster, Auslander, and Allen, 1976; Ipaktchi, et al., to appear). Our purpose here has been to illustrate a number of novel dynamical phenomena that have proven useful in interpreting experimental data on population systems.

2.9. While we have employed a 2-dimensional model for illustrative purposes, none of the dynamical behavior we have described is peculiar to this case. Similar bifurcation phenomena can be observed in models with continuous time and age structures. Indeed, the chaotic behavior characteristic of strange attractors is being seriously investigated as an explanation for the phenomenon of turbulence in fluid mechanics (where the bifurcation parameter for the Navier-Stokes equation is the Reynolds number).

In the context of population ecology the possibility that density dependent effects can produce chaotic appearing dynamics poses some interesting issues for modelling. It is probably not possible to collect field data of sufficient accuracy to employ nonlinear models as quantitatively predictive tools for populations where such effects are important. Therefore, it is necessary to reassess the sort of questions we ask of mathematical models. In many instances only qualitative or statistical information can be gleaned from a model. The "macroscopic" dynamical features we have stressed may be the final level of description for nonlinear models in such circumstances.

3. TWO POPULATION SYSTEMS

3.1. It is a simple matter to couple two single-population models together to study how competition or predation affects demographic behavior. Since this doubles the dimensionality of the model, as well as introducing additional parameters to model the interaction, it is apparent that the range of dynamic possibilities is concomitantly greater. Here we shall briefly illustrate how the bifurcation phenomena discussed above has helped to elucidate some aspects of an experimental host-parasite system.

The experiments we are referring to were performed by C. B. Huffaker and his associates on the moth *Anagasta kühniella* and the parasitic wasp *Venturia canescens* (Huffaker and Stinner, 1971; Hassell and Huffaker, 1969), cf. Figure 11. They found that, under a variety of laboratory conditions, the two populations could coexist for long periods—several years in one case. However, both populations showed regular oscillations, and the phase synchrony between the two populations appeared to govern the stability of the system. The age-specific character of the interaction between

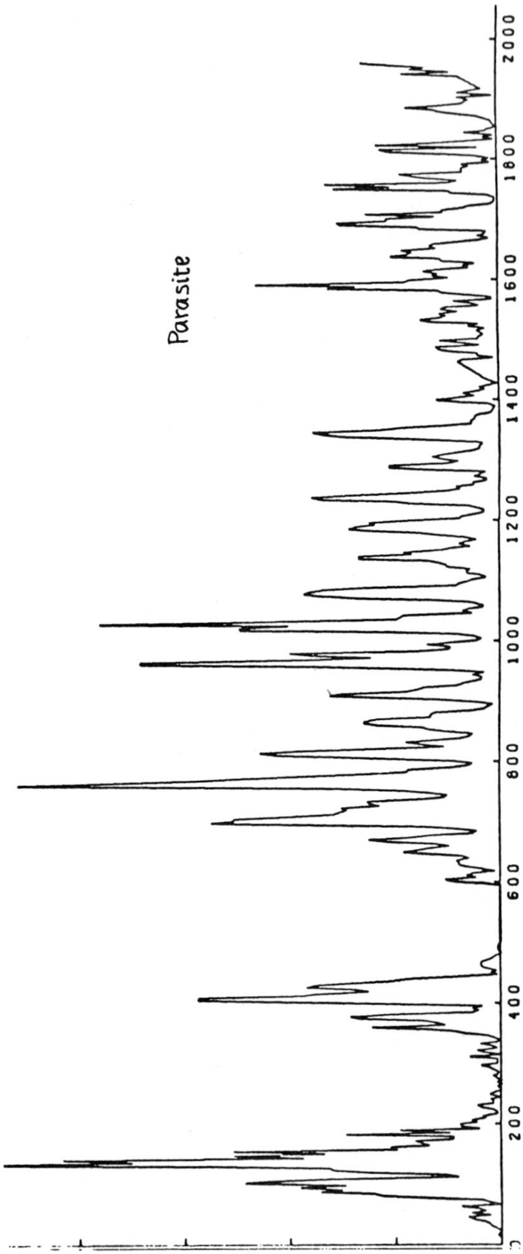
Parasite
200
400
600
800
1000
1200
1400
1600
1800
2000

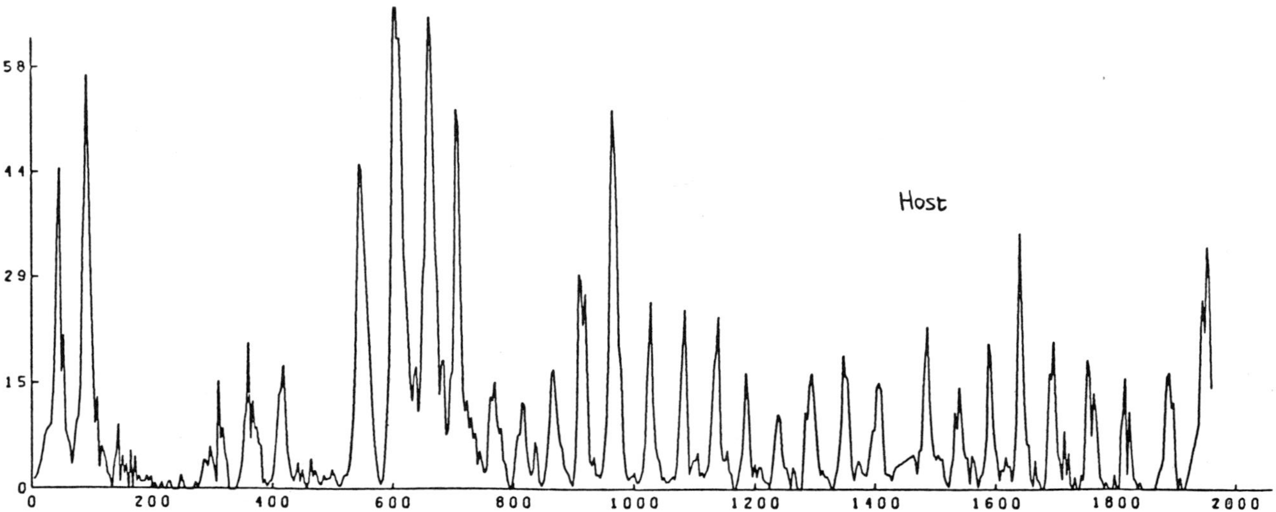
Host
58
44
29
15
0
0
200
400
600
800
1000
1200
1400
1600
1800
2000

FIG. 11.

the two populations turned out to be the crucial effect causing the populations to synchronize.

Before examining the age-structured model, however, let us first look at the simplest one-age-class model for a host-parasite system first formulated by Nicholson and Bailey (1935).

3.2. We assume that the host population grows logistically in the absence of the parasite: $H_{t+1} = H_t e^{r(1-H_t/k)}$ (cf. equation 6b, 11). If the parasite is assumed to search randomly over a fixed area, then one can show that the host population will be attenuated by a factor e^{-aP_t} = probability of escaping parasitization, where P_t is the parasite population in generation t and a is the "area of discovery" (Hassell and Varley, 1969). Thus, the host population is governed by

$$H_{t+1} = H_t e^{r\left(1-\frac{H_t}{K}\right)} \cdot e^{aP_t}. \tag{15}$$

The parasite population is therefore given by

$$P_{t+1} = \alpha H_t (1 - e^{-aP_t}), \tag{16}$$

where α is the number of ovipositions on each host.

We can reduce the number of parameters to two by nondimensionalizing. Define: $x_t \equiv H_t/K$, $y_t \equiv P_t/K_\alpha$, then (15) and (16) become

$$x_{t+1} = x_t e^{[r(1-x_t)-\gamma y_t]} \tag{17}$$

$$y_{t+1} = x_t(1 - e^{-\gamma y_t}) \tag{18}$$

where $\gamma \equiv Ka\alpha$.

A numerical study of this system was carried out by Beddington, *et al.* (1975). They found a progression of bifurcations similar to that discussed above for the 2-age class model. For their parameter set the initial bifurcation produced a set of invariant circles which gave way to a sequence of $5 \cdot 2^k$-point cycles (since the initial instability occurred near arg $(z) = 2\pi/5$). Eventually, a complex strange attractor made its appearance. Thus we see that, even for this simplest model, the host and parasite can coexist in one of 4 stable modes: (i) a time-invariant equilibrium, (ii) one of a large number of periodic orbits, (iii) an almost periodic orbit (invariant circle), (iv) a chaotic, or statistical, equilibrium in the strange attractor.

As we mentioned earlier, it is generally impossible in practice to distinguish long cycles from chaos. Moreover, it is an open question how—or whether—one can distinguish the "deterministic chaos" associated with a strange attractor from truly random behavior when stochastic effects dominate the system. Consequently, we shall focus our attention on the relationship between coexistence and periodicity in host-parasite systems. First, however, it is worth pointing out that natural selection could drive the system into the chaotic regime as a condition for coexistence. Although the model is rather crude, the form of the equations (17) and (18) suggests how this might come about. The parameter r is the host population's net rate of increase and γ measures the parasites' ability to locate and reproduce in the host larvae. Together, r and γ determine which of the four stable coexistence modes the system inhabits. That is, the parameter plane (r, γ) can be partitioned into disjoint subregions, each characterized by either an equilibrium, cycles or chaos (some regions also correspond to $P_t = 0$ or $(H_t, P_t) = 0$). Associated with each point (r, γ) is an asymptotic population level $[\bar{x}(r, \gamma), \bar{y}(r, \gamma)]$, where $(\bar{x}, \bar{y})$ are the equilibrium values in region (i) or in regions (ii), (iii), and (iv), a time average $(1/\tau)\Sigma_{j=0}^{\tau} x_{k-j}$ over the past τ generations. Thus we can superimpose on the (r, γ)-plane two sets of level lines: $\bar{x}$ = constant and $\bar{y}$ = constant. This is shown schematically in Figure 12. If we regard r and γ as "strategic parameters" for the host and parasite, respectively, then the situation is very similar to a continuous, 2-person game (Ho, 1970). The competitive equilibrium in parameter space (r^*, γ^*) is the so-called "Nash point", n, from which neither "player" can deviate without decreasing his utility—in this case the average population. Because of the asymmetry in the bifurcation diagram it turns out that n can easily lie in the cyclic or chaotic region, indicating that each population can maximize its average level by adopting the "unpredictable" demographic strategy characteristic of a strange attractor.

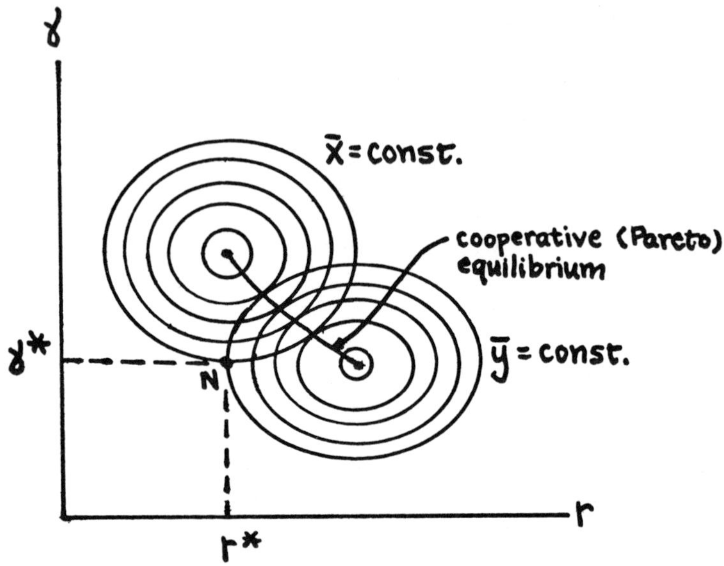

FIG. 12.

Of course, the situation is much more complicated than this. In particular, we have said nothing about the genetic constraints imposed on the evolution of the parameters r and γ. It turns out that, when equations (17) and (18) are augmented by the simplest models for Mendelian inheritance, the demographic equilibrium point may be incompatible with the genetic equilibrium! (Rocklin and Oster, 1976). Thus, the system can never achieve a steady state, but must traverse a periodic or aperiodic orbit indefinitely. This is a complicated issue which we shall defer to a later study and return to the consideration of age-structure effects.

3.3. The simplest generalization of model (15), (16) is to include two age classes in each population as shown schematically in Figure 13. The equation of motion for this system is

$$\mathbf{F}:\mathbf{R}^2 \to \mathbf{R}^2,$$

$$(P_1, P_2, H_1, H_2) \mapsto \big(H_1(1 - e^{-aP_2}),$$

$$SP_1, H_2 e^{r(1-H_2/k)}, H_1 e^{-aP_2}\big). \tag{19}$$

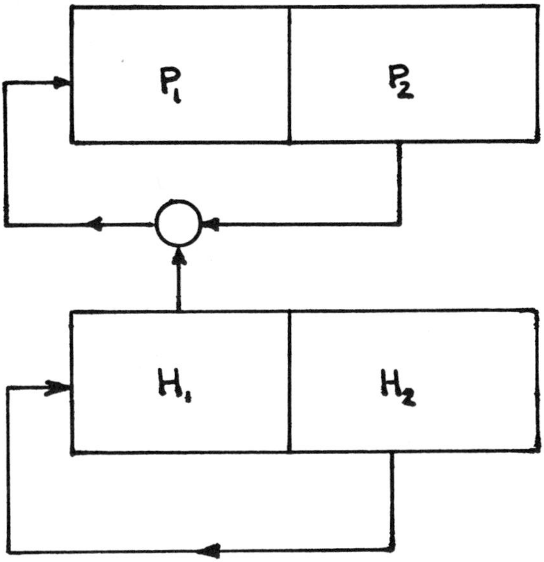

FIG. 13.

The system is 4-dimensional now; however, the dynamics can be best appreciated by examining the interacting age classes H_1 and P_2. Numerical simulations show that the populations can easily coexist in a number of periodic as well as chaotic states. Unfortunately, the high frequency oscillations which appear to dominate the periodic orbits here are an artifact of the small number of age classes. We can smooth the trajectories considerably by adding more age classes. In Figure 14 we show a simulation of the host-parasite system which includes some 50 age classes. This model was simulated using the actual data from the experiments cited in section 3–1. A complete discussion of the model can be found in Auslander, Oster and Huffaker (1974). From Figure 14 the mechanism by which the two populations coexist can be understood. When the adult parasite population is at its peak the host larvae upon which it preys is safely ensconced in other, invulnerable, age classes. Conversely, when the host larvae are at their peak levels there are few adult parasites. An examination of the complete age profiles of each population reveals a pair of "travelling waves" which propagate through the age structures. The "phase" between these waves are displaced by such an amount that the parasite is forced to live off the "tails" of the host wave, as shown in Figure 14. Thus, although the experimental system was quite homogeneous spatially, by adopting a particular dynamic age structure, the host was able to establish a "temporal niche" in which it was relatively free from predation. Conversely, the parasite was able to pursue the host to some extent by adopting a wave-like age structure which could "pursue" the host population waves. Numerical studies on the model show that this coexistence mode is quite stable, and independent experimental confirmation of its reality was obtained by Podoler (1974).

One further property of the model deserves comment. When the parameters are such that the systemse such that the system is in the quasi-periodic mode (invariant circle) then the "beat" phenomenon can work to the advantage of the host population. The mean generation time of the host and parasite populations are generally not the same (25 days for the parasite and 51 days for the host in Huffaker's experiments). Thus, if the host larvae remain at a low level for a length of time equal to the adult parasite lifespan an entire generation of parasites will find no oviposition opportunities, and consequently the parasite will be wiped out and the host can then achieve a much higher average population level. This can also happen if the peak population of the parasite is too large and

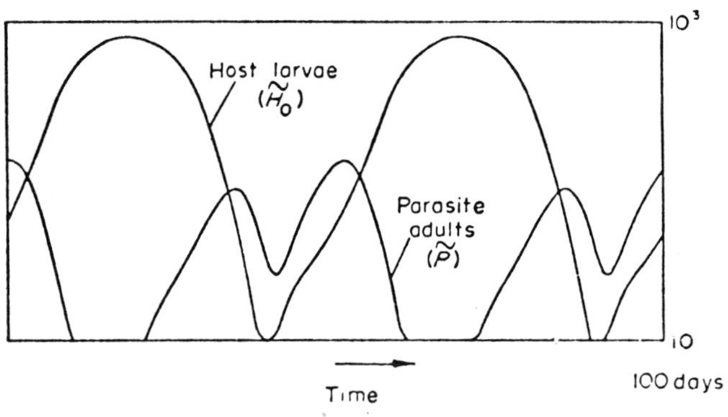

FIG. 14.

correctly synchronized with the host larvae, then the hosts will be driven to such a low level that elimination of the parasite is inevitable. Indeed, such behavior is observed in the model when in the "beat" mode, and has important implications for biological control programs which depend on parasite introduction.

REFERENCES

1. D. Auslander, G. Oster, and C. Huffaker, "Dynamics of interacting populations," *J. Franklin Inst.*, (5) **297** (1974), 345–376.
2. J. Beddington, C. Free, and J. Lawton, "Dynamic complexity in predator-prey models framed in difference equations," *Nature*, **255** (1975), 58–60.
3. J. Guckenheimer, G. Oster, and A. Ipaktchi, "Dynamics of density dependent population models," *Theor. Pop. Biol.* **4** (1977), 101–147.
4. M. Gurtin and R. MacCamy, "Non-linear age-dependent population dynamics," *Arch. Rational Mech. Anal.*, **54** (1974), 281–300.
5. M. Hassell and G. Varley, "New inductive population model for insect parasites and its bearing on biological control," *Nature*, **223** (1969), 1133–1137.
6. M. Hassell and C. Huffaker, "Regulatory processes and population cyclicity in laboratory populations of Anagasta kühniella (Zeller) (Lepidoptera: Phycitidae). III. The development of population models," *Res. Pop. Eco.*, **11** (1969), 186–210.
7. Y. C. Ho, "Differential games, dynamic optimization and generalized control theory," *J. Optimization Theory Appl.*, **6** (1970), 179–209.
8. C. Huffaker and R. Stinner, "The role of natural enemies in pest control programs," *Entomological Essays to Commemorate the Retirement of Professor K. Yasumatsu*, Hokuryukan Publ. Co., Tokyo, 1971, 333–350.
9. T-Y. Li and J. Yorke, "Period three implies chaos," *Amer. Math. Monthly*, **83** (1976), 985–992.
10. J. Marsden and M. McCracken, "The Hopf Bifurcation," *Applied Mathematical Sciences*, vol. 19, Springer-Verlag, New York, 1976.
11. R. M. May, "Biological populations with nonoverlapping generations: stable points, stable cycles, and chaos," *Science*, **186** (1974), 645–647.
12. R. May and G. Oster, "Bifurcations and dynamic complexity in simple ecological models," *Amer. Natural.*, **110** (1976), 573–599.
13. N. Metropolis, M. Stein, and P. Stein, "On finite limit sets for transformations on the unit interval," *J. Combinatorial Theory*, **15** (1) (1973), 25–43.
14. A. Nicholson and V. Bailey, "The balance of animal populations, Part I," *Proc. Zool. Soc. London* (1975), 551–598.

15. A. J. Nicholson, "The self-adjustment of populations to change," *Cold Spring Harbor Symposia on Quantitative Biology*, **22** (1957), 153-173.

16. G. Oster, D. Auslander, and T. Allen, "Deterministic and stochastic effects in population dynamics," *Trans. ASME*: *J. Dyn. Sys., Meas. and Control*, **98** (1976), 44–48.

17. G. Oster, "Lectures in population dynamics; Modern Modelling of Continuum Phenomena," R. DiPrima, ed., *Amer. Math. Soc.*, 1976.

18. H. Podoler, "Effects of intraspecific competition in the Indian meal moth on the populations of the moth and its parasite *Nemeritis canescens*," *J. Anim. Ecol.*, **43** (1974), 641–651.

19. ———, "Analysis of life tables for a host and parasite (Plodia-nemeritis) ecosystem," *J. Anim. Ecol.*, **43** (1974), 653–670.

20. S. Rocklin and G. Oster, "Competition between phenotypes," *J. Math. Biol.*, **3** (1976), 225–261.

21. S. Smale and R. Williams, "The qualitative analysis of a difference equation of population growth," **3** (1976), 1–4.

22. K. Swick, "A nonlinear age-dependent model of a single species population dynamics," *SIAM J. Math. Anal.*, 1975.

POPULATION MODELS AND COMMUNITY STRUCTURE IN HETEROGENEOUS ENVIRONMENTS

Simon A. Levin

INTRODUCTION

Until recently, the bulk of the extensive mathematical literature in ecology ignored the role of spatial heterogeneity. There were exceptions, for example the classic work of Skellam (1951); but despite much earlier developments in the related areas of epidemics (Brownlee 1911) and genetics (Fisher 1937, Haldane 1948), it has only been in the last few years that theoretical studies in ecology have begun to take account of the fundamental importance of geographic distribution. Recent reviews of the subject may be found in Levin (1976a, 1976b).

In part, this current attention to spatial factors may be traced to parallel developments in the ecological experimental literature, especially island colonization studies (Schoener 1975; Simberloff 1976; Simberloff and Wilson 1969, 1970; Sutherland 1974) and the associated theory of island biogeography (MacArthur and Wilson 1967). Even ostensibly non-insular environments, such as forests, farms, or rocky shores, may often be viewed as mosaics of islands of different characteristics (Levin and Paine 1974, 1975), in essence archipelagoes of microhabitats (Cromartie 1975, Janzen 1968, Platt 1975, Root and Chaplin 1976).

Further, Hutchinson (1951) emphasized long ago the influence of heterogeneity on adaptive strategies, and there has been growing interest in recent years in colonization strategies, especially dispersal (Den Boer 1968, 1971; Gadgil 1971; Reddingius 1971; Smith 1972; Strathman 1974) and dormancy (D. Cohen 1970, Marks 1974), and their relation to the spatial and spatio-temporal characteristics of the environment. Simulation studies have been undertaken to shed some light on these evolutionary questions (Allen 1975; Caswell 1976; Reddingius and Den Boer 1970; Roff 1974a, 1974b, 1975), but analytical results are nearly nonexistent.

The second major factor contributing to the development of the mathematical theory of spatially distributed populations is the increasing theoretical interest in the prototype mathematical description, the diffusion-reaction system (Fife 1976a, 1976b; Greenberg and Hastings 1977; Kopell and Howard 1974; Othmer 1975, 1977; Othmer and Scriven 1971, 1974; Torre 1975), especially in biological contexts (Aronson and Weinberger 1975, 1977; Dubois and Adam 1976; Fife and Peletier 1977; Hadeler 1977; Keller and Segel 1970; Kolmogoroff, Petrovskij, and Piskunov 1937; Segel and Jackson 1972; Segel and Levin 1976; Smale 1974; Steele 1974a, 1974b). A more complete review may be found in Levin (1976b).

The consideration of spatial effects fundamentally changes our view of the organization of ecological communities. This paper is an attempt to survey the most important implications.

BASIC MATHEMATICAL APPROACH

Although numerous alternatives exist and may sometimes be mandated (Levin 1976b), the most usual description of spatially distributed populations is through a discrete or continuous version of a transport-reaction system. Although movement is not generally of purely diffusive nature, such systems are often referred to generically, and somewhat sloppily, as *diffusion-reaction* systems even when non-diffusive terms are included. These systems provide the most convenient starting point for a discussion of the mathematical approach, and I shall first discuss them and some of the ways that they have found usage in ecology.

The discrete formulation is a bit more flexible with regard to the transport schemes which can be conveniently accommodated, and assumes an environment consisting of m distinct cells (numbered $\mu = 1, \ldots, m$) immersed in a much larger "bath." For each i, the dynamics of the density u_i^μ of species i in patch μ are assumed to be governed by an equation of the form

$$du_i^\mu / dt = f_i(\mathbf{u}^\mu, \mathbf{v}^\mu) + \{\text{net exchange with other patches}\}$$
$$+ \{\text{net exchange with bath}\}, \tag{1}$$

in which $\mathbf{u}^\mu$ is the vector $(u_1^\mu, \ldots, u_n^\mu)$ of densities in cell μ and $\mathbf{v}^\mu$ is the corresponding vector of environmental parameters (e.g., altitude, temperature, moisture) and may be time dependent. The functions f_i^μ are arbitrary except that they are assumed to be defined and continuously differentiable on an open set the projection of which onto "$\mathbf{u}^\mu$ – space" contains the set $R_\mu = \{\mathbf{u}^\mu : u_i^\mu \geqslant 0 \forall i\}$; and further, for $\mathbf{u}^\mu \varepsilon R_\mu$, $f_i^\mu \geqslant 0$ if $u_i^\mu = 0$ (Levin 1976a). The function f_i^μ, the localized growth rate for species i in patch μ, depends only on local densities and parameters. The assumption made above simply says that this localized growth rate cannot be negative if species i is not present. Some special cases of (1) will be considered later.

Many and perhaps most environments are truly patchy, and the discrete formulation (1) provides a better description than does a continuous one. However, a continuous model is more amenable to analytical treatment, and there are many situations when the continuous approximation may be justified. Indeed, for some (such as aquatic systems) it is probably to be preferred on biological grounds.

In the continuous version, the cell mosaic is replaced by a continuum, and each point identified by its spatial coordinates $\mathbf{x}$. In the simplest versions, movement is assumed to be by passive diffusion, although the diffusion coefficients may vary in the different directions (as will be discussed later). The result is a system of nonlinearly coupled partial differential equations of parabolic type for the species' densities. When the habitat is linear and only one spatial dimension is to be considered, these equations take the form

$$\partial u_i/\partial t = f_i(\mathbf{u}, \mathbf{v}) + \partial(D_i \partial u_i/\partial x)/\partial x, \tag{2}$$

in which D_i is the diffusion coefficient for species i and might depend on x, u, and t. The extension to two dimensions is obvious, yielding

$$\partial u_i/\partial t = f_i(\mathbf{u}, \mathbf{v}) + \partial\left(D_i^1 \partial u_i/\partial x_1\right)/\partial x_1 + \partial\left(D_i^2 \partial u_i/\partial x_2\right)/\partial x_2, \tag{3}$$

where D_i^1 and D_i^2 are the diffusion coefficients in the two directions. If $D_i^1 = D_i^2 = D_i$ and D_i is constant, (3) degenerates to the familiar form

$$\partial u_i/\partial t = f_i(\mathbf{u}, \mathbf{v}) + D_i \nabla^2 u_i. \tag{4}$$

A slightly more general movement regime introduces advection into the system for instance due to water current velocity $w(t)$. In the one-dimensional case, this replaces (2) by

$$\partial u_i/\partial t = f_i(\mathbf{u}, \mathbf{v}) + \partial(D_i \partial u_i/\partial x)/\partial x - w(t)\partial u_i/\partial x. \tag{5}$$

Platt and Denman (1975), in one of the most detailed efforts to provide a mathematical model for plankton distribution, utilize a three-space-dimensional version

$$\partial b/\partial t = f(b, \mathbf{v}) + D_H \nabla_H^2 b + D_{x_3} \partial^2 b/\partial x_3^2 - \mathbf{w} \cdot \nabla b, \tag{6}$$

in which b is phytoplankton density, $\mathbf{w}$ is the water velocity vector, D_H and D_{x_3} are horizontal and vertical diffusion coefficients, and $f(b, \mathbf{v})$ has the special form

$$f(b, \mathbf{v}) = k_m\left[i_{x_3}/\left(K_i + i_{x_3}\right)\right]b - R_m\left[1 - e^{-\lambda(b-b_0)}\right]^+. \tag{7}$$

In this particular formulation, the first term represents growth where k_m is the maximal cell division rate (at optimal light intensity), $i_{x_3} = i_0 \exp(-\alpha x_3)$ is the availability of light at depth x_3 (α is an optical extinction coefficient), and K_i measures the strength of the dependence of cell division on light. The second term is the loss due to herbivory and is of a type due to Ivlev; R_m is the maximal feeding rate, b_0 is a food threshold level, and λ is a constant. The notation $[\ ]^+$ indicates that the value within the bracket appears in (7) if it is positive, but is replaced by zero if it is negative.

The particular form of (6) reflects the fact that the diffusion coefficients in the two horizontal directions may be taken to be equal, but the coefficient in the vertical direction will in general be quite different.

Equation (6) is an equation for one species only (or more correctly, for the sum of all phytoplankton species). Of course, in reality, the feeding rate must depend on zooplankton density (since these are the principal herbivores); if the particular form (7) is accepted, this means that R_m must depend on zooplankton density. Since zooplankton density itself will be related to phytoplankton density, principally through effects upon rate of change of its density, (6) must at the very least be replaced by a system of two equations of predator-prey type. I shall return to discuss this problem further in the next section.

ECOLOGICAL APPLICATIONS OF THE BASIC APPROACH

There is considerable literature, much of recent vintage (see for example Aronson and Weinberger 1975, 1977; Fife 1976a, 1976b; Fife and Peletier 1977; Greenberg and Hastings 1977; Hadeler 1977; Kolmogoroff, Petrovskij, and Piskunov 1937; Kopell and Howard 1974; Othmer 1975, 1977; Segel and Levin 1976; Torre 1975) on the properties of the solutions to diffusion-reaction systems of the form (4), and a somewhat smaller literature (see for example Levin 1974, 1976a; Othmer and Scriven 1971, 1974; Segel and Levin 1976) on the discrete version (1). Their interest is generally in the mechanisms which lead to the formation of spatial or spatio-temporal patterns, including waves, wave fronts, and pulses (Aronson and Weinberger 1975: Fife 1976a, 1976b; Greenberg and Hastings 1977; Hadeler 1977; Othmer 1977; Segel and Levin 1976). No attempt will be made to survey those mechanisms here; rather, I shall illustrate by examples how these influence thinking about ecological communities.

1. SPREAD OF A POPULATION RELEASED AT A SINGLE POINT

This is one of the simplest of applications and has appeared often in the ecological literature (see for example Skellam 1951). Growth is considered negligible, and assuming the spread is isotropic and two-dimensional with constant rate of spread the population is modeled by an equation of the form

$$\partial u/\partial t = D(\partial^2 u/\partial x^2 + \partial^2 u/\partial y^2) = D\triangle^2 u. \tag{8}$$

The one- and three-dimensional versions show no significantly different behavior. Equation (8) is a two-dimensional version of (5), but with $f_i = w(t) = 0$. The solution, subject to the specified initial conditions that the release took place at the origin, takes the well-known form of a normal distribution

$$u(x, y, t) = (c/t)\exp\left[-(x^2 + y^2)/4Dt\right], \tag{9}$$

in which c is a constant. Note that the variance of this distribution is $\sigma^2 = 2Dt$, which increases linearly with time. This formula (9) changes slightly if diffusion is not isotropic or if the number of dimensions is changed, but the key property that the variance increases linearly with time remains true.

If the release does not take place at a single point, or if one simply wishes to predict the future dispersion of the population given the distribution at time zero, then one obtains the initial value problem

$$\begin{aligned} \partial u/\partial t &= D(\partial^2 u/\partial x^2 + \partial^2 u/\partial y^2), \\ u(x, y, 0) &= f(x, y). \end{aligned} \tag{10}$$

The solution of this is well known, being constructed by superposition of fundamental solutions of the form (9):

$$u(x, y, t) = (1/4\pi t)\int_{-\infty}^{\infty}\int_{-\infty}^{\infty} f(\xi, \eta)$$
$$\times \exp - \left(\left[(x-\xi)^2 + (y-\eta)^2\right]/4Dt\right)d\xi d\eta. \qquad (11)$$

Note that this will reduce to the form (9) if f is a multiple of a δ-function centered at the origin, and further that

(i) The total population size, $U = \int\int u(x, y, t)dxdy$ is independent of time, and is thus equal to

$$\int_{-\infty}^{\infty}\int_{-\infty}^{\infty} f(\xi, \eta)d\xi d\eta.$$

This may be verified directly from (11), making appropriate changes in the order of integration and using the identity

$$1 = (1/4\pi t)\int_{-\infty}^{\infty}\int_{-\infty}^{\infty} \exp\left[-(x^2+y^2)/4Dt\right]dxdy.$$

However it simply reflects the fact that the underlying equations describe only the redistribution of the population, and not growth or decay.

(ii) The center of the population,

$$\mathbf{M} = (\bar{x}, \bar{y}) = \int_{-\infty}^{\infty}\int_{-\infty}^{\infty} (x, y)\cdot u(x, y, t)dxdy/U,$$

also is independent of time; and in fact $\bar{x} \equiv 0, \bar{y} \equiv 0$. This simply describes the fact that due to symmetry, there is no change in the population center.

(iii) The variance of (x, y),

$$V = \int_{-\infty}^{\infty}\int_{-\infty}^{\infty} (x^2+y^2)u(x, y, t)dxdy/U,$$

increases linearly with time according to the relation

$$V = V_0 + 2Dt.$$

Okubo (1970) has shown that the above model does not precisely conform to diffusion in the surface layers of oceans and lakes, but needs to be modified somewhat. Indeed, dye diffusion experiments show a quadratic or cubic dependence on t for the variance rather than the linear dependence of Fick's diffusion (above). His proposed modification replaces (8) by

$$\partial u/\partial t = (c/r)\partial/\partial r(r^m \partial u/\partial r) \tag{12}$$

in which c and m are constants and $r^2 = x^2 + y^2$. Equation (8) is a special case of (12) with $m = 1$, but Okubo's data places the correct value of m at approximately 2.7.

2. CHEMICAL COMMUNICATION

Animals within a species signal danger, resource location, and sexual availability by emitting chemicals known as pheromones. Rubinow (1973) and Bossert and Wilson (1963) discuss mathematical aspects of the spread of the alarm pheromone in the harvester ant, *Pogonomyrmex badius*, using a three-dimensional version of the basic diffusion approach described above, with an additional factor of 2 introduced in the solution because of reflection at the surface of the earth. (This is equivalent to imposing the boundary condition $\partial u/\partial z = 0$ at $z = 0$.) The basic formula (9) is then replaced by

$$u(x, y, z, t) = 2N/(4\pi Dt)^{3/2} \exp\left[-(x^2 + y^2 + z^2)/4Dt\right], \tag{13}$$

in which the point of release is considered to be virtually at the surface of the earth. Here, u is the pheromone concentration and N is the amount released at time zero.

Okubo (1977) discusses further developments in the theory,

considering continuous emission from a moving source and emission in a wind, thereby introducing advective effects and the complications of turbulent diffusion.

3. PLANKTONIC PATCHINESS

One of the earliest applications of the basic diffusion approach to ecological problems was due to Kierstead and Slobodkin, who were motivated by a need to understand the causes of the initiation of red tides, large blooms of certain phytoplankton (particularly dinoflagellates) which can be extremely harmful to fish populations. Qualitatively, what appears to trigger these blooms is a local accumulation of favorable nutrients and other requisite organic growth substances. The quantitative theory began with Kierstead and Slobodkin, who initially consider one horizontal dimension and introduce a simple logarithmic growth term into the one-dimensional version of (8). This results in the equation

$$\partial u/\partial t = D\partial^2 u/\partial x^2 + ru, \tag{14}$$

in which u is the phytoplankton concentration and r the "intrinsic rate of natural increase." The form (14) is equivalent to assuming a local growth rule of the form

$$du/dt = ru,$$

which would have as solution $u = u_0 \exp(rt)$; hence the name logarithmic or exponential growth. Although it is an oversimplified description of population growth and not in general valid, it may be justified in this case since interest is upon the problem of initiation of tides. Hence logarithmic growth may be regarded as simply a linear approximation near $u = 0$.

In an infinite region, the solution to (14) is obtained simply:

$$u(x, t) = \left[e^{rt}/(2\sqrt{\pi}\, t^{1/2})\right]\int_{-\infty}^{\infty} f(\xi)\exp\left[-(x-\xi)^2/4Dt\right]d\xi, \tag{15}$$

where $f(\xi)$ describes the initial data. It is easy to show, most simply by using the methods of section (1) for the case $r = 0$, that the total population size grows exponentially with exponent rt, the population center remains static, and the variance increases linearly with t.

However, the problem of interest to Kierstead and Slobodkin did not involve an infinite region. Rather, it is assumed that a (long narrow) favorable region of horizontal diameter L (and transverse cross-section area A) develops and is surrounded by a totally unfavorable region in which phytoplankton concentration cannot attain positive values; this introduces the boundary value problem

$$\partial u/\partial t = D\partial^2 u/\partial x^2 + ru, 0 < x < L, \qquad (16)$$
$$u(0, t) = u(L, t) = 0.$$

Roughly speaking, the population size will increase at a rate related to the "volume" of the region, which is proportional to L, minus the losses due to dissipation through the boundaries of the region, which are independent of L. If L is too small, the population cannot grow fast enough to compensate for those losses; but beyond a critical diameter L_c, there will be a net growth. This value L_c should be a lower bound to the size of blooms, and the diameter of the smallest favorable region which will permit blooms to arise.

The dependence of L_c on the parameters D and r can be roughly determined by dimensional arguments. The units of D are $(\text{length})^2/\text{time}$ and of r, $(\text{time})^{-1}$; and hence it is to be expected that

$$L_{cr} \propto \sqrt{D/r}\ .$$

This may be rigorously verified and the constant of proportionality determined by direct consideration of the boundary value problem (16), as I now demonstrate.

One way to find particular solutions of (14) is by separation of variables; that is, by looking for solutions of the form $A(x)B(t)$. When this is done, one finds the solutions $\exp[ikx + (r - k^2D)t]$, where k is an arbitrary real number known as the wave number. Since we are only interested in real solutions, we are restricted to the real and complex parts of these separable solutions; that is, $\cos kx \exp(r - k^2D)t$ and $\sin kx \exp(r - k^2D)t$. Further, and more important, one can establish using Fourier transform arguments that any solution to (14) may be constructed by (an infinite) superposition of such separable solutions and further without loss of generality restrict attention to wave numbers $k \geqslant 0$.

For the boundary value problem (16), however, not every wave number is available but only those which result in the boundary conditions being satisfied. Since $\sin kx = 0$ for $kx = 0$ and in general for kx any multiple of π, this means that $kL = n\pi$ for some $n > 0$. Thus the wave numbers k are limited to the "discrete spectrum" $n\pi/L$, $n = 1, 2, 3, \ldots,$ which is to say (since the cosine terms will not satisfy the boundary conditions) that the general solution to (14) is of the form of a convergent series

$$u(x, t) = \sum_{n=1}^{\infty} B_n \cdot \sin(n\pi x/L) \exp(r - n^2\pi^2D/L^2)t. \quad (17)$$

For each n for which $r < n^2\pi^2D/L^2$, the corresponding term in (17) will be damped to zero as $t \to \infty$. In particular, if $r < \pi^2D/L^2$, every term and hence $u(x, t)$ itself will be damped to zero as $t \to \infty$. This means that the phytoplankton population will die out unless $L^2 \geqslant \pi^2D/r$, and hence the critical diameter is given by

$$L_{cr} = \pi\sqrt{D/r}\ . \quad (18)$$

[The identical result, but in a somewhat different setting, was obtained by Skellam (1951).] Compare this result with the heuristic relationship determined by dimensional arguments alone.

This is, of course, only the simplest imaginable model, but the results turn out to be quite robust. Kierstead and Slobodkin extend it to two horizontal dimensions by assuming that the region of interest is a cylinder of diameter L and depth h, and assume cylindrical symmetry. Formula (18) is then replaced by

$$L_{cr} = 4.81\sqrt{D/r}\ , \quad (19)$$

which is seen to be only a slight modification of (18). Again, this parallels a result of Skellam (1951). Okubo (ms.) retains the cylindrical (that is to say, horizontally circular) form, but considers three different diffusion laws. These give different forms for the critical size problem; but on the basis of empirically estimated constants, all turn out to be of the same order of magnitude (1-10 km.).

Dubois (1975a, 1975b), Platt and Denman (1975), and Wroblewski, O'Brien, and Platt (1975) discuss certain minor modifications of the results of Kierstead and Slobodkin when interacting populations are considered. The latter two papers reinterpret the relation (18) by introducing a grazing term as in (7) and find only a small percent correction to the critical length-scale.

As mentioned earlier, although Platt and Denman (1975) introduce a grazing effect, they do not consider directly the dynamics of the grazing population. Various attempts to do this, however, have been undertaken by Dubois (1975a, 1975b), Levin and Segel (1976), Okubo (1974), and Steele (1973, 1974a, 1974b). Part of the motivation is that, as Platt and Denman (1975) write, "fluctuations in the spatial distribution of phytoplankton have been found on all length-scales so far examined in the sea." Steele (1975) argues that the larger scale patchiness cannot be related to variation in physical parameters such as salinity or temperature, and so an alternative explanation in terms of biological interactions is needed.

Steele (1974a), as do several other authors, begins from the classical Lotka-Volterra equations

$$\begin{aligned} dP/dt &= aP - bPH, \\ dH/dt &= cPH - dH, \end{aligned} \tag{20}$$

in which P is the phytoplankton concentration and H the concentration of herbivorous copepods (zooplankton). These equations, as is well known, exhibit rather peculiar behavior. Their solutions are closed orbits in the P-H plane, but are not limit cycles. Rather, (20) describes something much like a harmonic oscillator in that P and H will oscillate in a pattern determined by initial conditions. Slight perturbations from the closed orbits will not be corrected, but instead will transfer the solution to a new closed orbit. An unstable equilibrium exists at $P = H = 0$ and a second, more interesting, equilibrium at $P = c/d$, $H = a/b$. The latter one, which is enclosed by the other orbits, is really a degenerate cycle and is, like the others, neutrally stable. Thus the system is not structurally stable in that small modifications to (20) will result in qualitatively different behavior. Therefore they provide a shaky starting point for investigation.

To consider the effects of diffusion, Steele adds appropriate terms to (20) to produce the system

$$\begin{aligned} \partial P/\partial t &= aP - bPH + D_1\partial^2 P/\partial x^2, \\ \partial H/\partial t &= cPH - dH + D_2\partial^2 H/\partial x^2. \end{aligned} \tag{21}$$

He specifically assumes $D_1 = D_2$, but we need not. The system has a spatially uniform equilibrium at $P = d/c$, $H = a/b$; to study its stability with respect to small perturbations p_0, h_0, one sets $P = (d/c) + p$, $H = (a/b) + h$ and discards higher-order terms to obtain the linearized system

$$\begin{aligned} \partial p/\partial t &= (-bd/c)h + D_1\partial^2 p/\partial x^2, \\ \partial h/\partial t &= (ac/b)p + D_2\partial^2 h/\partial x^2. \end{aligned} \tag{22}$$

In particular, if one seeks solutions to (22) of the form $p = A(t)e^{ikx}$, $h = B(t)e^{ikx}$, one finds that $A(t) = A_0e^{\sigma t}$, $B(t) = B_0e^{\sigma t}$ where

$$\begin{pmatrix} -D_1k^2 & -bd/c \\ ac/b & -D_2k^2 \end{pmatrix}\begin{pmatrix} A_0 \\ B_0 \end{pmatrix} = \sigma\begin{pmatrix} A_0 \\ B_0 \end{pmatrix}. \tag{23}$$

σ is thus an eigenvalue of the matrix

$$M = \begin{pmatrix} -D_1k^2 & -bd/c \\ ac/b & -D_2k^2 \end{pmatrix},$$

and hence has negative real part. Thus all finite wavelength perturbations will be damped, and the system with diffusion is seen to be more stable than the original. This conclusion is not restricted to the special solutions determined above but is general since any solution may be Fourier synthesized from them.

Steele, however, (for a particular boundary-value problem) argues that this linear approach is inappropriate because, by neglecting higher-order (quadratic) terms in (22), one is missing the fact that small wavelength perturbations can interact to produce ones of longer wavelength. This complaint is inappropriately applied to the system (21), since the basic underlying system (20) is structurally unstable; Murray (1975) (considering the case $D_1 = D_2$) demonstrates the error. Steele (1974b), however, extends his considerations to the more general system

$$\begin{aligned} \partial P/\partial t &= aP - f_1(P)H + D_1\partial^2 P/\partial x^2, \\ \partial H/\partial t &= f_2(P)H - f_3(H) + D_2\partial^2 H/\partial x^2, \end{aligned} \tag{24}$$

in which $f_1(P)$ is the grazing rate of H on P, $f_2(P)$ is the growth rate for herbivores, and $f_3(H)$ is the predation rate on the herbivores. Again Steele assumes $D_1 = D_2$, but this is not necessary. He further assumes that (24) has a uniform equilibrium P_0, H_0, which will satisfy

$$\begin{aligned} aP_0 &= f_1(P_0)H, \\ f_2(P_0)H_0 &= f_3(H_0). \end{aligned} \tag{25}$$

If one linearizes about that equilibrium, one obtains

$$\begin{aligned} \partial p/\partial t &= \alpha p + \beta h + D_1\partial^2 p/\partial x^2, \\ \partial h/\partial t &= \gamma p + \delta h + D_2\partial^2 h/\partial x^2, \end{aligned} \tag{26}$$

where

$$\begin{aligned} \alpha = a - f_1'(P_0)H_0,\ \beta &= -f_1(P_0), \\ \gamma = f_2'(P_0)H_0,\ \delta &= f_2(P_0) - f_3'(H_0). \end{aligned} \tag{27}$$

Note that in general $\beta < 0$ and, since prey are good for predators,

$\gamma > 0$. Using the equilibrium relations (25), one may rewrite the formulas for α and δ as

$$\alpha = H_0[f_1(P_0)/P_0 - f_1'(P_0)], \delta = [f_3(H_0)/H_0 - f_3'(H_0)]. \tag{28}$$

Either may be legitimately expected to be positive, thereby introducing autocatalytic effects, since such forms of the predator "functional response" are well-known (Holling 1965, Oaten and Murdoch 1975). Indeed the Ivlev-type grazing function discussed earlier (7) is of the type which would result in such autocatalysis.

If α and δ are positive, or indeed if simply $\alpha + \delta > 0$, then the uniform equilibrium (25) is unstable in the absence of diffusion ($D_1 = D_2 = 0$). For general perturbations of the type considered previously, the stability matrix has the form

$$M = \begin{pmatrix} \alpha - D_1 k^2 & \beta \\ \gamma & \delta - D_2 k^2 \end{pmatrix}. \tag{29}$$

This can have eigenvalues with positive real part if and only if k is sufficiently small; in other words, there will be a critical value $k_c > 0$ such that perturbations with wave number $> k_c$ will be damped, while those with smaller wave number will grow. In this situation short-wavelength perturbations may interact nonlinearly to reinforce, in particular, ones of longer wavelength. This may cause equilibria which are stable to small perturbations to become unstable in the presence of sufficiently large ones. Thus the uniform distribution is destabilized and patchiness initiated.

Levin and Segel (1976) and Okubo (1974) advance another hypothesis; they suggest that patchiness may arise by a process known as "diffusive instability," in which a stable uniform state is destabilized by the differential dispersal (diffusive) abilities of the species. The basic idea derives from an earlier suggestion by Turing (1952) and first advanced in an ecological setting by Segel and Jackson (1972) and Levin (1974).

Levin and Segel's hypothesis is an extension of considerations introduced by Segel and Levin (1976), based on the system

$$\begin{aligned} \partial P/\partial t &= P(a + eP - bH) + D_1 \nabla^2 P, \\ \partial H/\partial t &= H(-d + cP - gH) + D_2 \nabla^2 H, \end{aligned} \tag{30}$$

in which all parameters are non-negative. (Segel and Levin 1976 also treat the discrete analogue, for which the basic conclusions are virtually identical.) The system may be regarded as obtained (in the non-diffusive case) by retention of the linear terms in Taylor expansions for $(dP/dt)/P$ and $(dH/dt)/H$. Since computation is near equilibrium, this is for most purposes justifiable. (Technically, for some nonlinear calculations at least the quadratic terms should be retained; but in what is described below these make no difference.) The positivity of e may be justified in the same manner as mentioned earlier, based on a grazing function which has the same general "diminishing returns" character as (7). With $D_1 = D_2 = 0$, the system (28) has a positive equilibrium

$$\hat{P} = (bd + ag)/(bc - eg),\ \hat{H} = (ac + de)/(bc - eg), \tag{31}$$

provided $bc > eg$, and it is stable provided $(ac + de)g > (bd + ag)e$. To simplify the arithmetic, Levin and Segel (1976) take $d = 0$, so that the second stability condition reduces to $c > e$. However, this does not change the qualitative conclusions and we treat here the more general case.

The uniform solution $P = \hat{P}$, $H = \hat{H}$ is also an equilibrium when D_1 and D_2 differ from zero, but that equilibrium is not necessarily stable. Consideration as before of disturbances proportional to $\exp(i\mathbf{k} \cdot \mathbf{x})$ gives rise to the stability matrix

$$\begin{pmatrix} e\hat{P} - D_1 q^2 & -b\hat{P} \\ c\hat{H} & -g\hat{H} - D_2 q^2 \end{pmatrix} \tag{32}$$

where $q = |\mathbf{k}|$. Provided

$$R = D_2/D_1 < R_{cr} = 1/\left[\sqrt{b/g} - \sqrt{(b/g) - (e/c)}\,\right]^2, \tag{33}$$

the matrix in (32) admits no eigenvalues with positive real part for any value of q. For $R > R_{cr}$, however, a real positive eigenvalue and hence destabilization will occur for certain values of q. Segel and Levin (1976) study nonlinearly the bifurcation from the uniform state (with $d = 0$) and show that the destabilized equilibrium is replaced by a spatially non-uniform steady state. Thus a pattern of patchiness can arise due to the effects of differential diffusive abilities in destabilizing a uniform pattern of phytoplankton and zooplankton. This adds yet another mechanism to those which may contribute to planktonic patchiness.

This mechanism for patch formation also found another and earlier application (Keller and Segel 1970) in the consideration of the initiation of aggregation in cellular slime molds.

4. COEXISTENCE IN HETEROGENEOUS ENVIRONMENTS

Levin (1974, 1976a, 1976b) develops a general theory of ecological interaction in heterogeneous environments based upon systems of the forms (1)-(4). One of the most important applications is to systems in which the underlying dynamics in the absence of diffusion admit multiple stable states. In numerous ecological situations, the outcome of competition (for example) is dependent on initial conditions (such as colonization episodes), or events which occur early in successional development. The existence of multiple basins of attraction means that the local environment can support alternative stable communities, and leads in patchy environments to increased resource partitioning and species diversity (Levin 1974, 1976a, 1976b; Whittaker and Levin 1977). In such environments, "overall system pattern is a mosaic of equilibrium patches, each slightly modified by some input from nearby patches" (Levin 1976b).

The proof that such pattern may result from multiple stable states is actually dependent on the following theorem (Levin 1974, 1976a):

THEOREM 1: *Assume that* $F_i^\mu(\mathbf{U}, \mathbf{D}) \geqslant 0$ *when* $u_i^\mu = 0$ *and* $\mathbf{U} > 0$, *and that the system* $du_i^\mu/dt = F_i^\mu(\mathbf{U}, \mathbf{D}_0)$ *has a stable equilibrium (in the sense that the eigenvalues of the Jacobian matrix have negative real parts) at* $\mathbf{U} = \mathbf{U}_0 > 0$. *Then for* $\mathbf{D}$ *sufficiently close to* $\mathbf{D}_0$, *the system* $du_i^\mu/dt = F_i^\mu(\mathbf{U}, \mathbf{D})$ *has a stable equilibrium at some point* $\mathbf{U}_\mathbf{D} > 0$, *where* $\mathbf{U}_\mathbf{D}$ *tends to* $\mathbf{U}_0$ *as* $\mathbf{D}$ *tends to* $\mathbf{D}_0$. *Here* $\mathbf{U} = (\mathbf{u}^1, \mathbf{u}^2, \ldots, \mathbf{u}^m)$, *where* $\mathbf{u}^\mu = (u_i^\mu, \ldots, u_n^\mu)$, *and* $\mathbf{D}$ *is a matrix of parameters. The notation* $\mathbf{U} > 0$ *means that all components of* $\mathbf{U}$ *are non-negative.* F_i^μ *is assumed continuously differentiable.*

Note that the notion of stability used is stricter than asymptotic stability in that the borderline case when eigenvalues have zero real part is not considered.

The proof of this theorem is given in Levin (1976a) as a special case of the following more general theorem (Theorem 2). A discrete time version was derived by Karlin and McGregor (1972).

THEOREM 2: *Let* $\mathbf{y} = (y_1, \ldots, y_p)$, $\mathbf{z} = (z_1, \ldots, z_q)$; *and for* $j = 1, \ldots, p$, *let* $G_j(\mathbf{y}, \mathbf{z})$ *be a continuously differentiable function of* $\mathbf{y}$ *and* $\mathbf{z}$ *and satisfy the inequality* $G_j(\mathbf{y}, \mathbf{z}) \geqslant 0$ *provided* $y_j = 0$ *and* $y \geqslant 0$. *Assume further that the system of differential equations*

$$\frac{dy_j}{dt} = G_j(\mathbf{y}, \mathbf{z}^\circ)$$

has a stable equilibrium at $\mathbf{y} = \mathbf{y}^\circ \geqslant 0$. *Then for* $\mathbf{z}$ *sufficiently close to* $\mathbf{z}^\circ$, *the system*

$$\frac{dy_j}{dt} = G_j(\mathbf{y}, \mathbf{z})$$

also has a stable equilibrium $\mathbf{y}(\mathbf{z}) \geqslant 0$, *and* $\mathbf{y}(\mathbf{z})$ *tends to* $\mathbf{y}^\circ$ *as* $\mathbf{z}$ *tends to* $\mathbf{z}^\circ$.

The assumption of stability of the original equilibrium is an essential ingredient not only for the stability of the perturbed non-negative equilibrium, but also for the existence of a feasible equilibrium. In particular, if the original equilibrium lay on a boundary plane ($y_i = 0$) and were unstable, then in general a perturbed equilibrium would exist but might have negative components.

As an application of the above theorem, consider the system of equations

$$du_i^\mu/dt = F_i^\mu(\mathbf{U}, \mathbf{D}) = f_i^\mu(\mathbf{u}^\mu) + \sum_{\nu=1}^{m} D_i^{\nu\mu}(u_i^\nu - u_i^\mu). \quad (34)$$

In this formulation as before u_i^μ is the density of species i in patch μ; but now, for simplicity, movement is assumed to be by passive diffusion, with $D_i^{\nu\mu}$ the exchange coefficient from patch ν to patch μ. The parameters $D_i^{\nu\mu}$ are positive, and $D_i^{\nu\mu} = D_i^{\mu\nu}$.

Suppose now that the system in patch μ, were it sealed off from the rest of the mosaic, would have a (non-negative) stable equilibrium at $\mathbf{u}_0^\mu$. Then in the absence of diffusion ($\mathbf{D} = \mathbf{0}$), the whole system has a stable equilibrium at $(\mathbf{u}_0^1, \ldots, \mathbf{u}_0^m)$; and the theorems guarantee that for slight diffusion (e.g., for $\mathbf{D} \neq \mathbf{0}$ but sufficiently close to 0) the system will still have a (non-negative) stable equilibrium which will be close to the equilibrium $(\mathbf{u}_0^1, \ldots, \mathbf{u}_0^m)$. In particular, if the individual equilibria $\mathbf{u}_0^\mu$ show differences from one another, the overall system will exhibit spatial heterogeneity. This heterogeneity might reflect local parameter differences which alter the local equilibrium states, but it might also occur when the patches represent identical habitats (all f_i^μ identical) provided the local dynamics (as represented by f_i^μ) admit more than one stable equilibrium. The extension of this and Theorems 1 and 2 to stable attractors is intuitively clear but has not been rigorously presented in the literature.

To specialize further, consider a network of m identical patches, in each of which considered alone one or more of the n species would become extinct but in each of which every species has the potential to survive given a sufficiently good beginning. Then, for any specified collection of species, there exists a number n such that, if the number of patches m exceeds n and if there is only slight migration between patches, a stable configuration is possible with all of those species present (Levin 1974). In particular, if the matrix $\mathbf{D}$ is irreducible, then species i will at equilibrium be represented in every patch.

These results are exemplified by consideration of the admittedly over-simplified Lotka-Volterra competition equations:

$$\begin{aligned} du_1/dt &= u_1(r - au_1 - bu_2), \\ du_2/dt &= u_2(s - cu_1 - du_2), \end{aligned} \tag{35}$$

where all coefficients are positive.

In an environment consisting of two patches, two competitors whose local dynamics are governed by (35) will be described by the equations

$$\begin{aligned} du_1^\mu/dt &= u_1^\mu(r - au_1^\mu - bu_2^\mu) + D_1(u_1^\nu - u_1^\mu), \\ du_2^\mu/dt &= u_2^\mu(s - cu_1^\mu - du_2^\mu) + D_2(u_2^\nu - u_2^\mu), \end{aligned} \tag{36}$$

where $\mu \neq \nu$ and $\mu, \nu = 1, 2$. Assume D_1 and D_2 are positive, but close to zero. Then if $r/s > \max(b/d, a/c)$, (36) has only one stable equilibrium, $u_1^1 = u_1^2 = r/a$, $u_2^1 = u_2^2 = 0$; further, every trajectory beginning in the positive orthant will tend to this equilibrium. If $r/s < \min(b/d, a/c)$, every such trajectory tends to the stable equilibrium $u_1^1 = u_1^2 = 0$, $u_2^1 = u_2^2 = s/d$. Hence, these cases always result in competitive exclusion of one species or the other.

Assume now r/s is intermediate between b/d and a/c. If $a/c > b/d$ (intraspecific competition outweighs interspecific), then (35) has a stable equilibrium with both species present, and this translates to the equilibrium [in (35)]

$$\begin{aligned} u_1^1 = u_1^2 &= (rd - sb)/(ad - bc), \\ u_2^1 = u_2^2 &= (sa - rc)/(ad - bc). \end{aligned} \tag{37}$$

It is non-negative, and is the only stable equilibrium for (36). As in the previous two cases, it is spatially homogeneous, and every trajectory originating in the positive orthant tends to it. (For a more general treatment, see Hastings 1977c. For the system (36) one can easily write down a Liapunov function valid for the whole region.) Thus, in all of the three cases so far considered, the system tends asymptotically to a spatially homogeneous one.

One case remains, however, defined by the inequalities $b/d > r/s > a/c$ (interspecific competition outweighs intraspecific). In this case, the system (35) admits two stable equilibria,

$$u_1 = r/a, u_2 = 0 \tag{38}$$

and

$$u_1 = 0, u_2 = s/d. \tag{39}$$

In other words, the two species cannot coexist. The system (36), however, admits not only the two homogeneous equilibria corresponding to (38) and (39), but also two heterogeneous ones in which in one patch u_1 and u_2 are approximately given by

$$u_1 = r/a + O(D_1 + D_2), u_2 = O(D_1 + D_2) > 0, \tag{40}$$

and in the other by

$$u_1 = O(D_1 + D_2) > 0, u_2 = s/d + O(D_1 + D_2). \tag{41}$$

To illustrate this more clearly, suppose for simplicity that $a = d$, $b = c, r = s, D_1 = D_2 = D$, and $b > a$. Then an equilibrium exists at

$$\begin{aligned} u_1^1 = u_2^2 &= (r - 2D)/2a \\ &\quad + \sqrt{(r - 2D)[r - 2D(b + a)/(b - a)]}\,/2a, \\ u_1^2 = u_2^1 &= (r - 2D)/2a \\ &\quad - \sqrt{(r - 2D)[r - 2D(b + a)/(b - a)]}\,/2a, \end{aligned} \tag{42}$$

provided D is not too large $[D \leqslant r(b - a)/2(b + a)]$. The stability of this equilibrium is determined by the eigenvalues of the matrix $I_2 \otimes J + DK \otimes K$, where

$$I_2 = \begin{pmatrix} 1 & 0 \\ 0 & 1 \end{pmatrix}, \quad K = \begin{pmatrix} 0 & 1 \\ 1 & 0 \end{pmatrix}, \quad \text{and}$$

$$J = \begin{pmatrix} r - 2au_1^1 - bu_2^1 - D & -bu_1^1 \\ -bu_2^1 & r - bu_1^1 - 2au_2^1 - D \end{pmatrix},$$

(Levin 1974), or equivalently (Friedman 1956) by the eigenvalues of the matrices $J \pm DK$. It is straightforward to show (Levin 1974) that the equilibrium is locally stable provided $D \leqslant r(b-a)/2(2b+a)$; the global behavior of the system remains an unsettled question. It is not surprising that the heterogeneous equilibrium destabilizes for D large, since in this case mixing is so complete that no heterogeneous pattern can be maintained. This example shows that coexistence is possible in a patchy environment between species which would otherwise exclude one another; that is, species may avoid competitive exclusion by habitat partitioning. Why this occurs may be understood in another way by writing (following Peter Yodzis, unpublished) equations for the mean densities of u_1 and u_2,

$$\bar{u}_1 = (u_1^1 + u_1^2)/2, \bar{u}_2 = (u_2^1 + u_2^2)/2.$$

Using (36) in the simple case just considered, one gets (after rearrangement)

$$\begin{aligned} d\bar{u}_1/dt &= \bar{u}_1(r - a\bar{u}_1 - b\bar{u}_1) - a\delta^2_{u_1} - b\,\mathrm{cov}(u_1u_2), \\ d\bar{u}_2/dt &= \bar{u}_2(r - b\bar{u}_1 - a\bar{u}_2) - a\delta^2_{u_2} - b\,\mathrm{cov}(u_1u_2), \end{aligned} \tag{43}$$

in which $\delta^2_{u_1}$ and $\delta^2_{u_2}$ are respectively the variance of u_1 and u_2 about their means, and $\mathrm{cov}(u_1u_2)$ is the covariance. Coexistence has been made possible because the basic interaction, as described by (35), has been modified to account for the fact that the rates of growth of the populations are influenced by the dispersion patterns of the species over the patchwork environment. Were the populations uniformly distributed, then $\delta^2_{u_1}$, $\delta^2_{u_2}$, and $\mathrm{cov}(u_1u_2)$ would all equal zero and (43) would reduce to (35); indeed, this does define the behavior of (36) on the invariant two-dimensional subspace $u_1^1 = u_1^2$, $u_2^1 = u_2^2$. Naturally, (43) by itself is not a complete description of the system; two additional equations are needed to describe the rates of change of the dispersion patterns. The above analysis, however, shows that this can operate to stabilize a heterogeneous pattern.

This phenomenon is ecologically very important, because it

means that in a patchy environment, a complex and spatially heterogeneous pattern of species abundance can result either due to micro-variation in parameters or due to multiple steady states, and small amounts of migration will not destabilize such patterns. The determinants of the pattern are factors not included in the equations (1) or (34), such as extinction and colonization episodes. Often these could be accommodated in the framework (1) by allowing the equations to be stochastic, particularly colonization either from within the system or from the bath. However, as I shall discuss later, this may not be the best method for treating colonization and extinction, which often occur on a different time scale.

Although these factors also carry analogous implications in continuous environments, the precise mathematical results are more difficult to develop, and the patterns will in general not be spatially fixed. The relevant mathematical results are extremely difficult to obtain, but some sense of the problems may be realized by consideration of the equation

$$\partial u/\partial t = F(u) + \partial^2 u/\partial x^2 \tag{44}$$

which describes the dynamics of a single variable distributed over a linear habitat (the units have been chosen so that the diffusion coefficient is 1). Here u could represent the density of a single species, or it might represent the relative frequency of one species in a collection of species. Fisher (1937), in considering the spread of an advantageous allele, introduced an equation of the form (44) for the relative frequency of the allele, and the more general equation has come under extensive study in a variety of applications. In particular, Kolmogorov, Petrovskij, and Piskunov (1937), Aronson and Weinberger (1975, 1977), and Kanél (1962) have intensively studied (44) with special attention to the existence of wave fronts (see also Fife 1976a, Hadeler 1977, Rinzel 1975, 1977). For the cases when $F(u)$ has either two or three zeroes, the existence of such fronts has been demonstrated, waves which can in some cases transform a particular region from one (possibly stable) equilibrium state to another. Further, Fife (1977) has shown that for infinite regions or for finite ones with no-flux boundary conditions, no genuinely stable non-uniform stationary solutions to (44) exist. When more species are considered, these general results do not apply; stable patterns may exist as the case considered by Segel and Levin (1976) demonstrates. The latter example, however, was of a predator-prey system, and to my knowledge the question of the existence of such patterns in the competitive case is still open.

OTHER APPROACHES

In the preceding section, we examined a number of examples of the application of the basic "diffusion-reaction" approach to ecological problems. The litany was by no means exhaustive; we did not discuss in detail competition along environmental gradients, chemotactic or other directed movement, Skellam's computation of the steady-state distributions under logarithmic and logistic growth in a number of different habitats, or the potential ecological importance of periodic wave-train solutions such as those discussed by Kopell and Howard (1973, 1974). Instead, it is worthwhile to recognize that in a number of important situations, the previous description is inappropriate either in temporal or spatial scale; in these cases, a different approach will be necessary.

5. EQUILIBRIUM ISLAND BIOGEOGRAPHY AND REGIONAL PATTERNS

The theory of island biogeography was developed in the context of oceanic islands (MacArthur and Wilson 1967, Simberloff and Wilson 1969, 1970). However, it is in the consideration of mainland situations where the island approach may find its greatest application, since a large number of communities can, due to their patchy composition, be viewed as aggregates of patches or islands (Cromartie 1975, Janzen 1968, Levin and Paine 1974, Root 1973). In some of these situations, there will be no mainland in the usual sense, but there will be a "bath" which serves as the source of new colonists. In the marine intertidal, for example, "islands" appear as gaps in mussel beds and new colonists come from planktonic larvae. In forests, canopy gaps are colonized predominantly from "seed baths." In other situations, there may not even be any distinguished source of colonists; rather, the system is closed in that new colonists come from other patches. Moreover, it is often the case that colonists travel widely, and the short-range movement characteristic of the diffusion-reaction approach becomes inappropriate (except possibly on a faster time scale). Many species, for example, may undergo long-range dispersal to find new habitats at one stage in the life cycle, but engage in short-range, basically diffusive-type movements for feeding in between major dispersal episodes.

Given that the diffusive-type movement must be replaced by an alternative scheme, the problem still exists to specify what that should be. An intermediate solution, which retains features of both long-range and short-range movement, is utilized in the stepping-stone models of genetic correlation (Kimura and Weiss 1964, Weiss and Kimura 1965). A more extreme but probably often justified approach (MacArthur and Wilson 1967) assumes that all movement is long-range and that the composition of an island's immediate neighbors is inconsequential. This permits one to classify each island according to its intrinsic characteristics (e.g., size, exposure, distance from source, etc.), and to study colonization patterns in terms of these parameters.

As we saw in the previous section, those characteristics may not uniquely determine even the end state of the island's successional development; this theoretical conclusion has been borne out in numerous field studies of colonization (Root 1973; Simberloff and Wilson 1969, 1970; Sutherland 1974). Random events may play an important role in an island's colonization, and this could conceivably even apply to the recovery following such a major disturbance as a glaciation (Davis 1976). In this section, we restrict our attention to the so-called "equilibrium theory" of island biogeography (MacArthur and Wilson 1967, Simberloff and Wilson 1969, 1970), which places major importance on the long-term (asymptotic) species composition of an island following a disturbance. As we shall see later, transient phenomena are often very important and may not be ignored if disturbance is common. There are, however, situations where it is applicable, and it provides a convenient starting point.

The equations of the general form (1), which describe species interactions within an island and possibly some short-range movements, are assumed in the equilibrium theory to pertain to a very rapid time scale, so rapid that on the colonization-extinction time scale the events of (1) may be regarded as having run their course and reached some generalized form of equilibrium. That equilibrium will be upset only when some new disturbance is superimposed on the system.

To simplify the considerations, we assume first that the environment is homogeneous in that each island (patch) is identical as regards habitat characteristics and accessibility to new colonists. This assumption can be later relaxed. Each patch is assumed to have a finite number of steady-state compositions, and our interest is on the overall distribution of these states. The basic approach is exemplified by considering that each island has two possible stable states, A_1 and A_2, and one unstable one (the vacant state). We assume, however, that the perturbations which can destabilize the vacant state (i.e., colonization events) do not occur on the fast time scale. A_1 and A_2 may be characterized by a single species, for example sugar maple (A_1) and beech (A_2) (see, for example, Smith 1972, Whittaker and Levin 1977), or by complexes of species. Following Levin (1976a, 1976b), let P_1 = the number of patches in state A_1, P_2 = number in state A_2, and K = total number of patches, so that $K - P_1 - P_2$ are vacant. The basic model might then take the form

$$dP_1/dt = F_1(P_1, P_2, t),\ dP_2/dt = F_2(P_1, P_2, t), \qquad (45)$$

where the time dependence represents parametric variation (e.g., seasonal) in growth characteristics, disturbance, or colonization.

As a particular example, assume that K is constant and that the rate of replacement of vacant patches by state i patches is of the form $(m_i P_i + b_i) \cdot V$, where b_i represents the rate of "colonization" from the bath and $m_i P_i$ from other patches in the complex. Assume further that e_i is the rate that type i patches are recycled to "vacant," i.e., the rate at which they go extinct. Then, since

$V = K - P_1 - P_2$, (45) becomes

$$\begin{aligned} dP_1/dt &= (K - P_1 - P_2)(m_1P_1 + b_1) - e_1P_1, \\ dP_2/dt &= (K - P_1 - P_2)(m_2P_2 + b_2) - e_2P_2. \end{aligned} \tag{46}$$

This system will always (given positive initial conditions) tend to a stable balance between the two species or states, which is not surprising given that there is a perpetual source of new colonists.

Suppose now that the community is sealed off from the bath, so that $b_1 = b_2 = 0$. Then (46) reduces to the simpler form

$$\begin{aligned} dP_1/dt &= P_1[(m_1K - e_1) - m_1(P_1 + P_2)], \\ dP_2/dt &= P_2[(m_2K - e_2) - m_2(P_1 + P_2)]. \end{aligned} \tag{47}$$

Clearly, (47) can have no stable equilibrium; the "species" which wins is the one with the larger value of m/e (if both are identical, the balance is neutrally stable, and either A_1 or A_2 will random walk to extinction).

However, add now another level of complexity. Assume that A_1 and A_2 each has the potential to invade the territory of the other and instantaneously displace it (although this form of invasion competition may be allowed to be very one-sided). Suppose, indeed, that A_1 transforms to A_2 at the rate $c_2P_1P_2$, and A_2 to A_1 at the rate $c_1P_1P_2$. In each case this means that the rate of invasion is proportional to both the number of invasible patches and the potential number of invaders. Invasion is a form of growth, being at no cost to the host patch which is assumed to have colonists to spare. (47) then becomes

$$\begin{aligned} dP_1/dt &= P_1[(m_1K - e_1) - m_1P_1 - (m_1 + c_2 - c_1)P_2], \\ dP_2/dt &= P_2[(m_2K - e_2) - (m_2 + c_1 - c_2)P_1 - m_2P_2], \end{aligned} \tag{48}$$

which is of the same general form as the Lotka-Volterra equations introduced earlier (35). Clearly, if $m_1K < e_1$, species A_1 disappears from the system; and if $m_2K < e_2$, species A_2 does likewise.

Assume therefore that for both species the critical m/e ratio exceeds $1/K$. (Note that the importance of this inequality is reflected in the fact that either species in isolation would reach an equilibrium in which it occupied $K - e/m$ of the patches.) Then species A_1 will survive in the system provided

$$\frac{K - e_1/m_1}{K - e_2/m_2} > 1 + \frac{c_2 - c_1}{m_1}, \tag{49}$$

and species A_2 provided

$$\frac{K - e_2/m_2}{K - e_1/m_1} > 1 + \frac{c_1 - c_2}{m_2}. \tag{50}$$

If the species were equal competitors, so that $c_1 = c_2$, one of these conditions would have to be violated and at least one species go extinct. If, however, say species A_1 is the superior competitor, so that $c = c_1 - c_2 > 0$, then coexistence is possible if species A_2 has a larger m/e ratio than does species A_1, but not so much larger that species A_1 is outcompeted. If coexistence results, it is because a balance has been reached between one species which is a superior competitor and the other which is a superior colonizer, an idea which goes back at least to Hutchinson (1951) and Skellam (1951). The superior colonizer is sometimes referred to as a "fugitive" or "opportunistic" species.

Levins and Culver (1971) wished to extend this notion to permit consideration of an intermediate state in which both species are present, and Horn and MacArthur (1972) do the same for an environment made up of two types of patches. Although this is of course possible to do (see Cohen 1970, Levin 1974, Slatkin 1974), the original models of Levins and Culver and of Horn and MacArthur are not correct (Levin 1974). As mentioned these ideas can obviously be extended to consideration of multiple patch types. This we shall not pursue here.

Hastings (1977a) analyzes a predator-prey system in a patchwork environment, basically implementing a system simulated by Caswell (1976). The approach is much like the one described in this section, except that invasion of a "prey" patch by a predator results in the local extinction of both species after a suitable delay time. The system he studies is thus a differential-delay system, and interesting results are obtained concerning the survival of the system. Extension to three species (one predator, two prey) was simulated by Caswell (1976), and Hastings (1977b) also provides some preliminary discussion of this problem.

6. NONEQUILIBRIUM ISLAND BIOGEOGRAPHY

In the last examples of the preceding subsection, we have technically deviated from a classical equilibrium view by allowing "mutations" from one equilibrium state to another, but this involves no fundamental change in our approach. In Levin (1976a, 1976b), the notion of steady state is extended to include the *plateau*, a long-lived state which nonetheless will at some point be replaced by a successor state. In both Levin (1976b) and Whittaker and Levin (1977), models are considered for a variety of successional schemes, both obligate and non-obligate, in which the successional stages are considered as plateaus. Within this view, indeed, it is appropriate to view even the most advanced ("climax") stage as a plateau, since it will in time be recycled through the earlier stages of succession. Indeed, because the later successional species are often the most susceptible to major disturbance (such as fire) and the early successional are best adapted to colonize following disturbance, ecological communities in many cases tend to have a built-in successional microcycle with all stages being ephemeral.

The definition of a "successional stage" is a somewhat arbitrary convenience, since ecosystem succession, especially in terms of such parameters as diversity, biomass, and nutrient accumulation, is basically a continuous process. However, the general approach described above, a view of succession as a procession of plateaus, can obviously be extended to any number of such plateaus. In the limit, as one approaches the successional continuum, the models assume a form similar to that employed by Levin and Paine (1974, 1975). Their general approach is to assume that community organization is determined by the local properties and relative frequencies of the *patches* (or *islands* or *gaps*) which major disturbance creates, and by the biological development of the individual patch. These patches may be the canopy gaps which tree falls create, the phytoplankton blooms created by local bursts of nutrients, or the holes in mussel beds created by wave damage (numerous other examples are given in Levin 1976b).

In the model of Levin and Paine, each patch is characterized at time t by its successional age a and by a vector ξ of other characteristics (such as size, species composition, humus or nutrient level). Within a patch, obviously, $da/dt = 1$; but further, the vector ξ changes according to the equation

$$d\xi/dt = g(t, a, \xi), \tag{51}$$

where g is the vector of mean growth rates of the components of ξ. The deceptively simple form of (51) may describe dynamics as straightforward as the rate of growth or shrinkage of the area of the patch, or as elusive as the interactive dynamics of component species. In the simplest versions (Levin and Paine 1974, 1975), ξ is a scalar measuring area.

Primary interest focuses upon the distribution of patches with respect to a and ξ at time t, defined by the density function $n(t, a, \xi)$. To describe the rate of change of n over time, one may use the same basic equation of von Foerster (1959) that Oster and Rubinow utilize elsewhere in this volume. Specifically, let $\mu(t, a, \xi)$ be the instantaneous extinction rate for patches of age a and size ξ. Then the dynamics of n are governed by the first-order equation

$$\partial n/\partial t + \partial n/\partial a + \operatorname{div}(n\mathbf{g}) = -\mu n. \tag{52}$$

(This can be extended to allow for the variance about the mean growth rate $\mathbf{g}$ by the introduction of a second-order term in (52). For a discussion, see the paper by Oster in this volume.)

Given the initial distribution $n(0, a, \xi)$ and the rate of new patch formation $n(t, 0, \xi)$, equation (52) may in general be solved in closed form by the method of characteristics. When ξ is a scalar, this solution is given (for $t \geqslant a$) in Levin and Paine (1974). As an example, when μ is constant and $g(\xi) = c\xi$, the complete solution is

$$n(t, a, \xi) = \begin{cases} n(t-a, 0, \xi e^{-ca}) \cdot \exp[-(c+\mu)a], \; t \geqslant a, \\ n(0, a-t, \xi e^{-ct}) \cdot \exp[-(c+\mu)t], \; t < a. \end{cases} \tag{53}$$

Although solutions of this form are technically correct, they may not suffice in that the "birth" function $n(t, 0, \xi)$, which influences the solution, is in many applications dependent upon the whole distribution $n(t, a, \xi)$ at time t. One technique is to obtain a solution in a form analogous to (53), and then substitute this back into the renewal equation relating $n(t, 0, \xi)$ and the rest of the distribution. This approach is discussed in some detail by Oster and Takahashi (1974) and Rubinow (1973).

By solving the equation (52) subject to the appropriate boundary and initial conditions, one obtains an expression for the distribution n in relation to indicators of disturbance. In this way, for example, one can determine the degree of patchiness and successional age distribution of a region in relation to such habitat parameters as degree of exposure to waves (in the case of the intertidal) or wind (in the case of forests, where blowdowns are important), etc. Once this distribution is known, species composition of the community may be inferred and such macroscopic parameters as diversity determined and related to disturbance. Thus, for example, if disturbance is low, one would expect the successional age distribution to be such that late successional species would be predominant and diversity low. If disturbance is high, early successional species would be in great abundance and diversity again low. For intermediate disturbances, however, the greatest mixture of species and the greatest diversity will be achieved.

This approach provides a powerful tool for dealing with spatial heterogeneity, quite distinct from the diffusion-reaction paradigm. It has been applied already to the intertidal and to forest dynamics, and offers great potential in a variety of situations both for the treatment of theoretical questions and as a tool for management, in which for example disturbance regimes (as represented say by lumbering) are to be manipulated. Further, it provides a framework which is well suited to the study of such adaptations as seed dormancy and dispersal (Levin, Cohen, and Hastings 1977), which are evolutionary responses to environments unpredictable in space and time.

SUMMARY

Until recently, mathematical ecology has paid little attention to the importance of spatial heterogeneity, and thus most models have ignored species dispersion patterns. This, however, has rendered the conclusions of such models of limited interest in an understanding of community structure, since spatial heterogeneity and the potential for dispersal fundamentally alter species interactions and dynamics and increase by several orders of magnitude the potential for coexistence. The classical approach, localized in space and asymptotic in time, focuses attention only on the competitive dominants and their associated predators and parasites. However, consideration of spatial extent not only allows for the coexistence of incompatible co-dominants, it also allows for spatio-temporal patterns of occupancy and death or emigration which translate into survival for ephemerals which occupy early or intermediate stages on the successional gradient.

In this paper a variety of models are considered, from those of the diffusion-reaction type to the "patch" models of more recent vintage, and applications to a variety of ecological situations are discussed.

REFERENCES

1. J. C. Allen, "Mathematical models of species interactions in time and space," *Am. Nat.,* **109** (1975), 319–342.
2. D. G. Aronson and H. F. Weinberger, "Nonlinear diffusion in population genetics, combustion, and nerve propagation," Partial Differential Equations and Related Topics, *Lecture Notes in Mathematics,* **446**, J. A. Goldstein, ed., Springer-Verlag, Heidelberg, 1975.
3. D. G. Aronson and H. F. Weinberger, "Multidimensional nonlinear diffusion arising in population genetics," *Advances in Math.*, 1977, (to appear).

4. W. H. Bossert and E. O. Wilson, "The analysis of olfactory communication among animals," *J. Theoret. Biol.,* **5** (1963), 443–469.

5. J. Brownlee, "The mathematical theory of random migration and epidemic distribution," *Proc. Roy. Soc. Edinburgh,* **31** (1911), 262–289.

6. H. Caswell, "Predator mediated coexistence of prey," manuscript, 1976.

7. D. Cohen, "A theoretical model for the optimal timing of diapause," *Am. Nat.*, **104** (1970), 389–400.

8. J. E. Cohen, "A Markov contingency table model for replicated Lotka-Volterra systems near equilibrium," *Am. Nat.*, **104** (1970), 547–559.

9. W. J. Cromartie, Jr., "The effect of stand size and vegetational background on the colonization of cruciferous plants by herbivorous insects," *J. Appl. Ecol.*, **12** (1975), 517–533.

10. M. B. Davis, "Pleistocene biogeography of temperate deciduous forests," *Geosci. and Man,* **13** (1976), 13–36.

11. P. J. Den Boer, "Spreading of risk and stabilization of animal numbers," *Acta Biotheor. Leiden,* **18** (1968), 165–194.

12. ———, "Stabilization of animal numbers and the heterogeneity of the environment: The problem of persistence of sparse populations," *Proc. Adv. Study Inst. Dyn. Numbers Popul. Oosterbeek,* 1970, (1971), 77–97.

13. D. M. Dubois, "Simulation of the spatial structuration of a patch of prey-predator plankton populations in the Southern Bight of the North Sea," *Proc. Liège Colloq. Ocean Hydrodyn. 6th Mém. Soc. Roy. Sci. Liège,* VII (1975a), 75–82.

14. ———, "A model of patchiness for prey-predator plankton populations," *Ecol. Model.*, **1** (1975b), 67–80.

15. D. M. Dubois and Y. Adam, "Spatial structuration of diffusive prey-predator biological populations: Simulation of the horizontal distribution of plankton in the North Sea," *System Simulation in Water Resources,* G. C. Vansteenkiste, ed., North Holland Amsterdam, 1976, 343–356.

16. P. C. Fife, "Singular perturbation and wave front techniques in reaction-diffusion problems," *Proc. AMS-SIAM Symp.,* New York, Amer. Math. Soc., Providence, 1976a.

17. ———, "Pattern formation in reacting and diffusing systems," *J. Chem. Phys.,* **64** (1976b), 554–564.

18. ———, "Stationary patterns for reaction-diffusion equations," *Nonlinear Diffusion,* Pitman, London, 1977 (to appear).

19. P. C. Fife and L. A. Peletier, "Nonlinear diffusion in population genetics," *Arch. Rational Mech. Anal.,* 1977, (to appear).

20. R. A. Fisher, "The wave of advance of advantageous genes," *Ann. Eugen. London,* **7** (1937), 355–369.

21. B. Friedman, "An abstract formulation of the method of separation of variables," *Proc. of the Conference on Differential Equations,* J. B. Diaz and L. E. Payne, eds., University of Maryland Bookstore, College Park, Maryland, 1956, 209–226.

22. M. Gadgil, "Dispersal: Population consequences and evolution," *Ecology*, **52** (1971), 253–260.

23. J. M. Greenberg and S. P. Hastings, "Spatial patterns for discrete models of diffusion in excitable media," *SIAM J. Appl. Math.*, 1977, (in press).

24. K. P. Hadeler, "Nonlinear diffusion equations in biology," Proc. of Conference on Partial Differential Equations, Dundee, Scotland, 1976, *Lecture Notes in Mathematics*, Springer-Verlag, Heidelberg, 1977.

25. J. B. S. Haldane, "The theory of a cline," *J. Genet.*, **48** (1948), 277–284.

26. A. Hastings, "Spatial heterogeneity and the stability of predator-prey systems," *Theor. Pop. Biol.*, **12** (1977a), 37–48.

27. ———, "Spatial heterogeneity and the stability of predator-prey systems: Predator mediated coexistence of prey." Manuscript, 1977b.

28. ———, "Models in population biology," Ph.D. Thesis, Cornell University, 1977c.

29. C. S. Holling, "The functional response of predators to prey density and its role in mimicry and population regulation," *Mem. Entom. Soc. Can.*, **45** (1965), 1–60.

30. H. S. Horn and R. H. MacArthur, "Competition among fugitive species in a harlequin environment," *Ecology,* **53** (1972), 749–752.

31. G. E. Hutchinson, "Copepodology for the ornithologist," *Ecology,* **32** (1951), 571–577.

32. D. H. Janzen, "Host plants as islands in evolutionary and contemporary time," *Am. Nat.,* **102** (1968), 592–595.

33. I. Kanél, "Stabilization of solutions of the Cauchy problem for equations encountered in combustion theory," *Math. Sb.,* **59** (1962), 245–288 (Suppl.).

34. S. Karlin and J. McGregor, "Polymorphisms for genetic and ecological systems with weak coupling," *Theor. Pop. Biol.*, **3** (1972), 210–238.

35. E. F. Keller and L. A. Segel, "Initiation of slime mold aggregation viewed as an instability," *J. Theoret. Biol.,* **26** (1970), 399–415.

35a. H. Kierstead and L. B. Slobodkin, "The size of water masses containing plankton blooms," *J. Mar. Res.*, **12** (1953), 141–147.

36. M. Kimura and G. H. Weiss, "The stepping stone model of population structure and the decrease in genetic correlation with distance," *Genetics,* **49** (1964), 561–576.

37. A. Kolmogoroff, I. Petrovskij, and N. Piskunov, "Étude de l'équation de la diffusion avec croissance de la quantité de matière et son application à un problème biologique," *Bull. Univ. Moscow Ser. Intl. Sect. A1,* **6** (1937), 1–25.

38. N. Kopell and L. N. Howard, "Plane wave solutions to reaction-diffusion equations," *Stud. Appl. Math.*, **52** (1973), 291–328.

39. ———, Pattern formation in the Belousov reaction, *Some Mathematical Questions in Biology*, S. A. Levin, ed., **6** (1974), 201–217. Lectures on Mathematics in the Life Sciences, vol. 7, Amer. Math. Soc., Providence.

40. S. A. Levin, "Dispersion and population interactions," *Am. Nat.*, **108** (1974), 207–228.

41. ———, Spatial patterning and the structure of ecological communities, *Some Mathematical Questions in Biology,* S. A. Levin, ed., **7** (1976a), 1–36. Lectures on Mathematics in the Life Sciences, vol. 8, Amer. Math. Soc., Providence.

42. ———, "Population dynamic models in heterogeneous environments," *Annu. Rev. Ecol. Syst.*, **7**(1976b), 287–310.

43. S. A. Levin, D. Cohen, and A. Hastings, Manuscript, 1977.

44. S. A. Levin and R. T. Paine, "Disturbance, patch formation, and community structure," *Proc. Nat. Acad. Sci. USA,* **71** (1974), 2744–2747.

45. ———, "The role of disturbance in models of community structure, Ecosystem Analysis and Prediction," S. A. Levin, ed., *SIAM*, (1975), 56–67.

46. S. A. Levin and L. A. Segel, "Hypothesis for origin of planktonic patchiness," *Nature,* **259** (1976), (5545) 659.

47. R. Levins and D. Culver, "Regional coexistence of species and competition between rare species," *Proc. Nat. Acad. Sci. USA,* **68** (1971), 1246–1248.

48. R. H. MacArthur and E. O. Wilson, *The Theory of Island Biogeography,* Princeton University Press, Princeton, 1967, 203 pp.

49. P. L. Marks, "The role of pin cherry (Prunus pensylvanica L.) in the maintenance of stability in northern hardwood ecosystems," *Ecol. Monogr.,* **44** (1974), 73–88.

50. J. D. Murray, "Non-existence of wave solutions for the class of reaction-diffusion equations given by the Volterra Interacting-population Equations with Diffusion," *J. Theor. Biol.* **52** (1975), 459–469.

51. A. Oaten and W. W. Murdoch, "Functional response and stability in predator-prey systems," *Am. Nat.*, **109** (1975), 289–298.

52. A. Okubo, "Oceanic mixing," Tech. Rep. 62, Chesapeake Bay Inst., Johns Hopkins University, Baltimore, 1970.

53. ———, "Diffusion-induced instability in model ecosystems: Another possible explanation of patchiness," Tech. Rep. 86, Chesapeake Bay Inst., Johns Hopkins University, Baltimore, 1974.

54. ———, "The critical size of water masses containing plankton blooms under scale-dependent diffusion," Manuscript, 1976.

55. ———, Ecology and Diffusion, *Lecture Notes in Biomathematics*, Springer-Verlag, Heidelberg, 1977, (in press).

56. G. Oster and Y. Takahashi, "Models for age-specific interactions in a periodic environment," *Ecol. Monogr.*, **44** (1974), 483–501.

57. H. G. Othmer, "Nonlinear wave propagation in reacting systems," *J. Math. Biol.*, **2** (1975), 133–164.

58. ———," Current problems in pattern formation," *Lectures on Mathematics in the Life Sciences*, S. A. Levin, ed., vol. 9, Amer. Math. Soc., Providence, 1977, (in press).

59. H. G. Othmer and L. E. Scriven, "Instability and dynamic pattern in cellular networks," *J. Theor. Biol.*, **32** (1971), 507–537.

60. ———, "Nonlinear aspects of dynamic pattern in cellular networks," *J. Theor. Biol.*, **43** (1974), 83–112.

61. T. Platt and K. L. Denman, "A general equation for the mesoscale distribution of phytoplankton in the sea," *Mém. Soc. Roy. Sci. Liège*, **7** (1975), 31–42.

62. W. J. Platt, "The colonization and formation of equilibrium plant species associations on badger disturbances in a tall-grass prairie," *Ecol. Monogr.*, **45** (1975), 285–305.

63. J. Reddingius, "Gambling for existence: A discussion of some theoretical problems in animal population ecology," *Acta Theor. Suppl. Climum*, added to *Acta Biotheor.*, Brill, Leiden, **20** (1971), 208 pp.

64. J. Reddingius and P. J. Den Boer, "Simulation experiments illustrating stabilization of animal numbers by spreading of risk," *Oecologia*, **5** (1970), 240–248.

65. J. Rinzel, "Neutrally stable travelling wave solutions of nerve conduction equations," *J. Math. Biol.*, **2** (1975), 205–218.

66. ———, "Integration and propagation of neuroelectrical signals," *MAA Studies in Mathematical Biology*, vol. **15**: Cellular Behavior and the Development of Pattern, S. A. Levin, ed., Math. Assoc. of America, Washington, 1977.

67. D. A. Roff, "Spatial heterogeneity and the persistence of populations," *Oecologia*, **15** (1974a), 245–258.

68. ———, "The analysis of a population model demonstrating the importance of dispersal in a heterogeneous environment," *Oecologia*, **15** (1974b), 259–275.

69. ———, "Population dispersal and the evolution of dispersal in a heterogeneous environment," *Oecologia*, **19** (1975), 217–237.

70. R. B. Root, "Organization of a plant-arthropod association in simple and diverse habitats: The fauna of collards (*Brassica oleracea*)," *Ecol. Monogr.*, **43** (1973), 95–124.

71. R. B. Root and S. J. Chaplin, "The life styles of tropical milkweed bugs, *Oncopeltus* (Hemiptera, Lygaeidae) utilizing the same host," *Ecology*, **57** (1976), 132–140.

72. S. I. Rubinow, "Mathematical problems in the biological sciences," *SIAM*, 1973, 90 pp.

73. A. Schoener, "Experimental zoogeography: Colonization of marine mini-islands," *Am. Nat.*, **108** (1974), 715–738.

74. L. A. Segel and J. Jackson, "Dissipative structure: An explanation and an ecological example," *J. Theor. Biol.*, **37** (1972), 545–559.

75. L. A. Segel and S. A. Levin, "Application of nonlinear stability theory to the study of the effects of diffusion on predator-prey interactions," Topics in Statistical Mechanics and Biophysics: A Memorial to Julius L. Jackson, R. A. Piccirelli, ed., *Proc. AIP Conf.*, **27** (1976), 123–152.

76. D. S. Simberloff, "Experimental zoogeography of islands: Effects of island size," *Ecology,* **57** (1976), 629–648.

77. D. S. Simberloff and E. O. Wilson, "Experimental zoogeography of islands: The colonization of empty islands," *Ecology,* **50** (1969), 278–296.

78. ———, "Experimental zoogeography of islands: A two-year record of colonization," *Ecology*, **51** (1970), 934–937.

79. J. G. Skellam, "Random dispersal in theoretical population," *Biometrika,* **38** (1951), 196–218.

80. M. Slatkin, "Competition and regional coexistence," *Ecology*, **55** (1974), 128–134.

81. S. Smale, "A mathematical model of two cells via Turing's equation," *Some Mathematical Questions in Biology*, J. D. Cowan, ed., **5** (1974), 15–26. Lectures on Mathematics in the Life Sciences, vol. 6, Amer. Math. Soc., Providence.

82. F. E. Smith, "Spatial heterogeneity, stability, and diversity in ecosystems," in "Growth by Intussusception: Ecological Essays in Honor of G. Evelyn Hutchinson," *Trans. Conn. Acad. Arts Sci.*, **44** (1972), 309–335.

83. J. H. Steele, "Patchiness," *Coastal Upwelling and Ecosystems Anal. Newsl.*, **2** (1973), 3–7.

84. ———, "Spatial heterogeneity and population stability," *Nature,* **83** (1974a), 248.

85. ———, "Stability of plankton ecosystems," *Ecological Stability*, M. B. Usher and M. H. Williamson, eds., Chapman and Hall, London, 1974b, 179–191.

86. ———, *The Structure of Marine Ecosystems,* Harvard University Press, Cambridge, 1975.

87. R. Strathman, "The spread of sibling larvae of sedentary marine invertebrates," *Am. Nat.,* **108** (1974), 29–44.

88. J. P. Sutherland, "Multiple stable points in natural communities," *Am. Nat*, **108** (1974), 859–873.

89. V. Torre, "Synchronization of nonlinear biochemical oscillators coupled by diffusion," *Cybernetics*, **17** (1975), 137–144.

90. A. Turing, "The chemical basis of morphogenesis," *Philos. Trans. Roy. Soc. London Ser. B*, **237** (1952), 37–72.

91. H. von Foerster, "Some remarks on changing population," *The Kinetics of Cellular Proliferation,* F. Stohlman, Jr., ed., Stratton, New York, 1959, 382–407.

92. G. H. Weiss and M. Kimura, "A mathematical analysis of the stepping-stone model of genetic correlation," *J. Appl. Probability,* **2** (1965), 129–149.

93. R. H. Whittaker and S. A. Levin, "The role of mosaic phenomena in natural communities," *Theor. Pop. Biol.*, **12** (1977), 117–139.

94. J. S. Wroblewski, J. J. O'Brien, and T. Platt, "On the physical and biological scales of phytoplankton patchiness in the ocean," *Mém. Soc. Roy. Sci. Liège,* **7** (1975), 43–57.

THE NEUTRALIST-SELECTIONIST CONTROVERSY

W. J. Ewens

The essence of Darwin's theory of evolution by natural selection is that some types of individuals are in a sense "better" than others and are able to leave a disproportionately large number of offspring in the following generation. This leads to a rather different proportion of the "better" type in the offspring generation, compared to the parent generation, since the offspring are assumed to inherit the superiority their parents enjoyed. In turn, they will leave a disproportionately large number of offspring in the following generation, and so on in succeeding generations, until in the course of time marked changes in the frequencies of various types arise, to the extent that the superior types in effect have driven out the remaining types from the populations. The nature of the population is now different from, and in some sense "superior" to, the original population, and in this way an act of evolution in the population has occurred.

Because the precise nature of hereditary processes was unknown to him, Darwin's argument was on certain points necessarily vague and incomplete. The science of genetics has, in the meantime, led to a rather complete understanding of these points. We now know that the characteristics which individuals have are determined by their genetic constitution (together and interacting with the environment), and that the word "genotype" can more precisely replace the word "type" in the above argument. Thus we say that the fundamental unit process of evolution consists of the change over time in the genetic composition of a population: more specifically, in the replacement of an "inferior" gene in the population by a "superior" gene. The subject of theoretical population genetics concerns itself with the mathematical description of such a process, and on the whole has succeeded in accounting quantitatively for the observed facts of evolution in genetic terms, assuming selective, mutational and other parameters of the order presently observed in the real world.

This does not mean, of course, that selection must be the force determining the gene replacement procedures which have made up the course of evolution. For example, the above argument does not take account of the finite size of populations and the consequent possibility that stochastic effects may cause "superior" genes not to take over in the population. Indeed, the proposition has now been put forward that a high proportion of the gene replacements in evolutionary history were not determined by selection, but occurred rather by purely random processes acting on selectively equivalent genetic types. This theory, called the "non-Darwinian" or the "neutral alleles" theory, has caused considerable controversy in biological circles since it conflicts so clearly with the main cornerstone of evolutionary thought, namely, Darwinian natural selection.

In this expository article we will, first of all, introduce briefly as much genetic background as is necessary for the subsequent mathematical argument. Then an account will be given of the standard theory of genetic processes involving selection and mutation, and finally (and most importantly for us) a description will be given of certain aspects of the mathematics of the non-Darwinian theory.

GENETIC BACKGROUND

The characteristics any individual has are determined by a combination of environmental and genetic factors. So far as the genetic factors are concerned, these are determined by the collection of genes which the individual has. Genes are physical objects and lie, rather like beads on a string, along the chromosomes which appear in the cells of living things. In most species of interest to us, chromosomes occur in pairs, and to any gene on a chromosome there is a corresponding gene (in the corresponding location) on the homologous chromosome of the pair. We use the word "locus" to describe the position on the chromosome occupied by any pair of genes of interest to us.

Consider, then, the pair of genes at some locus on the chromosome. This pair will normally control, or partly control, one or more characteristics in the individual. In some cases there is almost a one-to-one relationship between a pair of genes and a given characteristic. For example, the nature of the genes at a certain locus determines whether or not the individual has sickle-cell anemia, in a way which we will make more precise later. Many disorders in humans are determined genetically in a similar fashion. On the other hand, some gene pairs partially control several characteristics, and a number of characteristics (for example, height) are controlled by a relatively large number of pairs of genes.

Returning to the pair of genes in which we are interested, we note that each gene may be of one or other of K types, which we label $A_1, A_2, \ldots, A_K$. The number K varies from locus to locus and in some cases can even be taken as infinity. There are, then, $K(K+1)/2$ possible types of individuals, namely A_1A_1, $A_1A_2, \ldots, A_KA_K$. The characteristic in question is determined by the genotype (for example, A_3A_4) of the individual in question. For example, at the sickle cell locus, there are two commonly occurring genes, denoted in this case by A (normal) and S (sickle). Individuals which are AA are normal and do not suffer from the disease, AS individuals have a very mild anemia, while SS individuals have severe anemia and do not usually survive to reproduce. This gives an example of differential selection acting on the genetic composition at a particular locus.

Before considering in general the effect of such differential selection on genetic evolution, we must consider the form of the frequencies of genotypes in any given generation. Suppose that at the time of conception of any generation the gene-types (or more properly, alleles) $A_1, \ldots, A_K$ at the locus of interest occur with frequencies $x_1, \ldots, x_K$, $(\Sigma x_i = 1)$, and that the population mates at random (i.e., the choice of mate is independent of the genetic make-up at the locus considered). Then supposing the size of the population to be indefinitely large, the frequency of A_iA_i at conception will clearly be x_i^2, while that of A_iA_j will be $2x_ix_j$. Frequencies of this nature are said to be in "Hardy-Weinberg" form (Hardy 1908, Weinberg 1908). Naturally these frequencies will change due to selective differentials during the growth to maturity of the generation in question. Nevertheless, the next generation will again have, at conception, genotype frequencies in Hardy-Weinberg form. Note, however, that the gene frequencies $x'_1, \ldots, x'_K$ in the next generation can, because of selection, be different from those in the parent generation.

How strong is such selection in real world cases, and how much will gene frequencies change in the course of one generation? The sickle cell case described above is somewhat unusual, first in that there is such a close correspondence between the genotype at the locus in question and a well-defined physiological phenomenon, and second in that the effect of genotype differences is so dramatic. For many other loci, the genotype occurring at the locus is not so important: the difference between having (say) A_2A_3 instead of A_4A_4 may be almost indistinguishable physiologically. Indeed, it is theoretically conceivable that there are loci where the genotype occurring is totally unimportant in that all possible genotypes have identical physiological effects in the organism. It is thus a matter of indifference to the individual which genotype he has at such a locus, and further, there can be no selective forces acting at such a "neutral" locus.

This does not mean that gene frequencies remain static in such a case. In a closed finite population, there are two other ways in which gene frequencies can change. First, a gene can mutate (i.e., spontaneously change its type). This will cause a change in the frequencies of the two types involved. Second, in finite populations stochastic effects normally ensure that even when no directed forces act, the genetic make-up of a daughter generation differs from that of the parent generation. This effect is relatively more important in a smaller population than in a larger one.

It is possible to conceive of cases where, at a selectively neutral locus, a new gene type arises by mutation and purely by chance increases in frequency to unity after perhaps a large number of generations. If we define as the fundamental unit of evolution the replacement of one gene type at a locus by another, then an act of evolution has taken place. However, we would hardly describe this as one which has "improved" the genetic nature of the population, and in the strict Darwinian sense we could say that evolution has not occurred. Organisms containing the new gene function physiologically in the same way as did their ancestors carrying the old gene.

The "non-Darwinian" or "neutral alleles" theory claims that a very high proportion of the genetic changes which have made up the course of biological evolution have been of this random, nonadaptive nature. As a corollary, it claims that a high proportion of the genetic variation seen in a population at any time point is also of this nonadaptive nature: we happen to see a population at a time point part way through the course of such a neutral evolutionary event.

The argument between those who adhere to the non-Darwinian theory and those who remain Darwinians is fierce and unresolved. The arguments on both sides often use, or were originally inspired by, mathematical analyses. We now review briefly the relevant mathematics for both theories.

SELECTION: THE STATIC THEORY

Consider a population of size so large that we may ignore stochastic fluctuations. We focus attention on a certain locus at which two gene types (or alleles), A_1 and A_2, can occur, and suppose that there exist selective differences between the three genotypes possible at this locus. Selection acts for two reasons, differential viability (i.e., survivorship to adulthood) and differential fecundity. The latter form of selection is relatively complex since it refers properly to mating pairs rather than single individuals, and to simplify the exposition we assume here that all selection is viability selection. Suppose, then, that the relative viabilities, or fitnesses, of the three genotypes are in the proportions shown below, and that genotype frequencies are in the Hardy-Weinberg form shown:

$$\begin{array}{llll} \text{genotype} & A_1A_1 & A_1A_2 & A_2A_2 \\ \text{fitness} & w_{11} & w_{12} & w_{22} \\ \text{frequency} & x_1^2 & 2x_1x_2 & x_2^2. \end{array} \qquad (1)$$

Clearly the frequencies of A_1 is x_1. What is the frequency x_1' of A_1 at the time of conception of the following generation? Elementary arguments show that

$$\begin{aligned} x_1' &\propto w_{11}x_1^2 + w_{12}x_1x_2 \\ 1 - x_1' &\propto w_{12}x_1x_2 + w_{22}x_2^2 . \end{aligned} \qquad (2)$$

From (2) we find that

$$x_1' = \{w_{11}x_1^2 + w_{12}x_1x_2\}/\overline{W}, \qquad (3)$$

where $\overline{W}$, the so-called mean fitness of the population, is defined by

$$\overline{W} = w_{11}x_1^2 + 2w_{12}x_1x_2 + w_{22}x_2^2. \qquad (4)$$

Selection of this sort acts in one or other of two ways on gene frequencies. Either one or other gene will be eventually driven out of the population ($x_1 \to 0$ or $x_1 \to 1$), or selection leads to a stable configuration (or polymorphism) where both genes coexist in the population. It is a classical result of population genetics theory (Fisher, 1922) that this latter event occurs if and only if w_{12} exceeds both w_{11} and w_{22}. We now consider elementary properties of such a polymorphism.

First, putting $x_1' = x_1$, equations (3) and (4) show that the equilibrium frequency x_1^* of A_1 is given by

$$x_1^* = (w_{12} - w_{22})/(2w_{12} - w_{11} - w_{22}). \qquad (5)$$

Note that $\overline{W}$ is maximized, as a function of x_1, at $x_1 = x_1^*$; the maximizing $\overline{W}$ is given by

$$\overline{W}^* = \left(w_{12}^2 - w_{11}w_{22}\right) / \left(2w_{12} - w_{11} - w_{22}\right). \tag{6}$$

Note further that since we normally compare fitnesses, the fact that the w_{ij} are relative rather than absolute values is not important; we do not pursue this matter here other than to note that in a population of stable size the convention is often adopted that $\overline{W}^* = 1$.

The sickle cell case is the classical example of such a stable polymorphism. It has been observed in certain African tribes that the "normal" gene A and the "sickling" gene S both occur in high frequencies. This arises because while the genotype SS seldom survives to maturity and (compared to unit fitness for the "normal" genotype AA) has fitness of essentially zero, the heterozygote AS has fitness in excess of that of AA because individuals of this genotype suffer only very mildly through sickling and appear to have increased resistance to malaria compared to AA individuals; their fitness is perhaps 1.25. Insertion of the relative fitnesses 1:1.25:0 into (5) would lead us to expect a frequency for the "normal" gene A to about .83, a value quite close to that actually observed.

The above considerations have not taken any account of the possibility of mutation; i.e., of the spontaneous change in the type of a gene from one form to another. Since mutation rates are usually very small (of order 10^{-5} at most), this approximation is often reasonable. However, mutation is a key factor in evolution and, further, the qualitative nature of the stochastic processes we describe later depends critically on whether or not mutation exists. Suppose then, in addition to the parameters given in (1) above, we assume that A_1 mutates to A_2 at rate u, and A_2 mutates to A_1 at rate v. Then equation (3) is amended to

$$x_1' = \left[(1-u)\left\{w_{11}x_1^2 + w_{12}x_1x_2\right\} + v\left\{w_{12}x_1x_2 + w_{22}x_2^2\right\}\right] / \overline{W}. \tag{7}$$

Assuming w_{12} exceeds both w_{11} and w_{22}, so that a polymorphism exists (in the present deterministic analysis) even without mutation, equation (5) must be replaced by a new equation yielding a new value for x_1^*. If u and v are small, the two expressions for x_1^* are usually extremely close, and negligible error arises in using equation (5) whether or not mutation exists.

We conclude this section by introducing the concept of the segregational genetic load. Suppose, for example, that the three genotypes A_1A_1, A_1A_2, A_2A_2 have fitnesses $1 - s$, $1 + s$, $1 - s$ respectively, and that mutation can be ignored. Equation (5) shows that the equilibrium frequency of A_1 is .5, and equation (6) shows that the mean fitness of the population is unity. Suppose now that a number n of loci, all with these fitness parameters, exist, and suppose further that the overall fitness of any individual is found by multiplying together the fitnesses of his genotypes at all n loci. Then the fitness of an individual heterozygous at all loci is $(1 + s)^n$ or approximately e^{ns} if n is large. For large n, this is a large value even for moderate s. Generally, it is clear that some individuals in the population are being asked to contribute considerably more offspring than average, and the population is said to be carrying a "segregational load".

We shall discuss this argument again later and note here simply that it was perhaps the main argument leading to the introduction of the neutral alleles theory.

SELECTION: THE DYNAMIC THEORY

Suppose the fitness of the three genotypes A_1A_1, A_1A_2, and A_2A_2 are such that $w_{11} > w_{12} > w_{22}$. Then equation (3) shows that $x_1' > x_1$, as we expect intuitively. As time goes on, the frequency of A_1 steadily approaches unity. Equation (3) shows that

$$x_1' - x_1 = x_1(1 - x_1) \times \left[w_{11}x_1 + w_{12}(1 - 2x_1) - w_{22}(1 - x_1)\right]\overline{W}^{-1}. \quad (8)$$

If we may assume $\overline{W} \cong 1$, we may approximate this equation by

$$\frac{dx_1}{dt} = x_1(1 - x_1)\left[w_{11}x_1 + w_{12}(1 - 2x_1) - w_{22}(1 - x_1)\right]. \quad (9)$$

Clearly the time required for the frequency of A_1 to increase from any value α to any higher value β is approximately

$$\int_{\alpha}^{\beta}\left[x(1-x)\{w_{11}x+w_{12}(1-2x)-w_{22}(1-x)\}\right]^{-1}dx \quad (10)$$

generations. Calculation of such times, in varying and complex sets of cases, were made in a sequence of important papers in the 1920's by Haldane (for summary, see Haldane 1932). For selective differentials of small size, such times can be extremely long, especially when the frequency of A_1 is close to unity. Indeed, formal application of (10) shows that the time required for the frequency of A_1 to reach unity is infinitely long. We shall see later how this conclusion is amended by considering the finite nature of the population size.

If substitutions occur at a number of loci simultaneously, the population is said to suffer a "substitutional load". This arises for the same reason as the segregational load; individuals with different genotypes will be required to contribute differentially to the next generation, and individuals of certain genotypes will be required to contribute considerably more offspring than average. The extent to which this is so will depend on the number of substitutions in progress at any time and on the selective differentials at the loci substituting. The calculations are more complex than those for the static segregational process, but the results are rather similar; if we assume multiplicative fitnesses, the substitutional load increases exponentially with both the selective differences and the number of loci undergoing substitution. Again, these arguments were among those used in the formulation of the neutral alleles theory.

SELECTION IN FINITE POPULATIONS

It has been assumed above that the population size is so large that stochastic factors can be ignored. The neutral theory relies fundamentally on the assumption of stochastic behavior, and thus we consider briefly the effect of a finite population size on the selective models we have been considering.

To do this it is necessary to set up a model of the stochastic process describing the formation of one generation from its predecessor. Several such models exist. We consider here only one example, the classical Wright-Fisher model (Wright 1931, Fisher 1930) of binomial (or more generally, multinomial) sampling. In brief, this model is as follows: suppose a locus admits alleles $A_1, \dots, A_K$ and suppose in any generation that the numbers of genes of these various allelic types are $N_1, \dots, N_K$. The population size N is assumed to be constant for all generations: note that $\Sigma N_i = 2N$, since each individual in the population carries two genes at the locus in question. Suppose under a deterministic theory (i.e., $N = \infty$) the relative frequencies of $A_1, \dots, A_K$ in the next generation, calculated by taking into account selection and mutation, are $x_1', \dots, x_K'$: note that x_i' will be a function of $N_1, \dots, N_K$ and the selection and mutation parameters. If the (random) numbers of $A_1, \dots, A_K$ genes in the following generation are denoted $N_1', \dots, N_K'$, then the model assumes that

$$Pr\{N_1' \cdots N_K' | N_1 \cdots N_K\} = \frac{(2N)!}{N_1'! \cdots N_K'!} \prod_{i=1}^{K} (x_i')^{N_i'}. \quad (11)$$

To exemplify the use of this model, suppose $K = 2$, that the selective parameters are as given in (1), and that A_1 mutates to A_2 at rate u and A_2 mutates to A_1 at rate v. Then x_1' will be given by equation (7) and thus equation (11) reduces to

$$Pr\{N_1' | N_1\} = \binom{2N}{N_1'} (x_1')^{N_1'} (1 - x_1')^{2N - N'_1}. \quad (12)$$

Note that it is sufficient to consider only the number N_1' of A_1 genes since N_2' is defined automatically by $N_2' = 2N - N_1'$.

Equation (12) defines a simple Markov chain describing the evolution of the frequency of A_1. It has the property that if $u = v = 0$, then eventually the frequency of A_1 will reach 0 or 1, irrespective of the values of the w_{ij}. On the other hand, if u and v are both positive, a stationary distribution for the frequency of A_1 will arise. If we make the approximation that x_1 is a continuous random variable in [0,1], and that the w_{ij} are all close to unity, the form of this stationary distribution, to a close approximation, is

$$f(x_1) = \text{const}\, x_1^{4Nv-1}(1-x_1)^{4Nu-1}\exp\{A(x_1)\}, \qquad (13)$$

where

$$A(x) = 2N\left\{w_{11}x^2 + 2w_{12}x(1-x) + w_{22}(1-x)^2\right\}.$$

It is of interest to compare the form of this distribution with the deterministic equilibrium point when w_{12} exceeds w_{11} and w_{22}, given to a sufficiently close approximation by (5). Suppose, for simplicity, that $u = v = 10^{-6}$, $N = 250{,}000$, $w_{11} = w_{22} = 1 - s$, $w_{12} = 1 + s$. Then in the deterministic case the frequency of A_1 is .5 irrespective of s, while in the stochastic case the stationary distribution of the frequency of A_1 is

$$f(x_1) = \exp\{2 \times 10^6 \times s x_1(1-x_1)\}, 0 \leqslant x_1 \leqslant 1. \qquad (14)$$

Clearly this distribution has mean .5, and the variance of the distribution will depend upon the value of s. Consider the three cases $s = 10^{-7}, 10^{-6}, 10^{-5}$. If $s = 10^{-7}$ the density function (14) is essentially uniform over [0,1]; for $s = 10^{-6}$ it exhibits a fairly well-defined maximum at $x_1 = .5$, although values of x_1 near 0 or which are not unlikely, while for $s = 10^{-5}$ substantial deviations of x_1 away from .5 are extremely unlikely. While one example can only give a rough indication of the effects of selection, mutation and population size on the stationary distribution, the general conclusion suggested by the above argument is in fact true: in a population of even moderate size, relatively minor selective forces can play a decisive role in the behavior of gene frequencies. This conclusion will be of importance to us when we consider the properties of the neutral theory.

Stochastic behavior also plays a role in the dynamic theory of gene substitution. Suppose mutation rates are zero and consider the stochastic analogue of equations such as (10). Here two questions may be asked. First, given the population size N, the fitness parameters w_{ij} and an initial frequency p of A_1, what is the probability that A_1 (rather than A_2) is fixed (i.e., reaches unit frequency) in the population? Second, how long may we expect such fixation to take?

The first question concerns a further point which leads to the neutral theory. Consider the case $w_{11} = w_{12} = w_{22}$, so that A_1 and A_2 are selectively neutral relative to each other. Arguments of symmetry show that the probability that A_1 eventually fixes is p. When the w_{ij} are no longer equal, the argument is much more complex. We do not enter into details other than to note that when $p = (2N)^{-1}$ (so that A_1 can be regarded as a new mutant in an otherwise pure A_2A_2 population), and when $w_{11} > w_{12} > w_{22} = 1$, the probability that A_1 eventually fixes is approximately

$$Pr\{A_1 \text{ fixes}\} \cong 2\{w_{12} - 1\}. \tag{15}$$

We return to this formula later. When the initial frequency of A_1 is not small, and the fitness parameters obey the inequalities just given, the probability that A_1 fixes is extremely close to unity. For example, when $N = 10^6$, $p = .5$, $w_{11} = 1.00002$, $w_{12} = 1.00001$, $w_{22} = 1.0$, the probability that A_1 eventually fixes is approximately .999999998. This again shows the "quasi-deterministic" behavior of a stochastic process with even quite small selective differentials.

The question of how much time such fixation requires is more complex, and we do not enter here into details. It is sufficient to mention that the very long times required under the deterministic theory for quite small changes in the frequency of A_1 (once this frequency is large) no longer apply; stochastic effects insure that the frequency of A_1 reaches unity far more rapidly than deterministic theory would suggest. For complete details on this point, see Ewens (1967).

POLYMORPHISMS

Most of the above theory was arrived at during the years before 1966. However, it had comparatively little interrelationship with real-world data. Little was known about "typical" values for selective differentials and population sizes, both crucial in determining the genetic evolution of a population. Much more important, however, comparatively little was known about the extent of genetic variation within natural populations and the extent of genetic differentiation between different populations and different species. Two developments led to a great increase of information on these two points. First, the technique of electrophoresis was applied, initially by Lewontin and Hubby (1966) in Drosophila and Harris (1966) in Man to determine the extent of genetic variation within populations. It was found that the amount of such variation was far larger than had previously been expected. Although information is still accumulating, it appears that perhaps 12% of the genome in any individual Drosophila is heterozygous, with an approximately similar figure in man and in a wide range of other species. This value is, if anything, probably an underestimate since the technique of electrophoresis is able to pick up only a fraction (perhaps 1/3) of actual genetic variation. On the other hand, the class of loci so far considered may not be typical of all loci on the chromosomes, and some care should be exercised in extrapolating present results.

When these results were known, Lewontin and Hubby immediately raised the question of the reasons for such widespread genetic variation. Several selective explanations are possible. First, as we have seen, the variation could arise by widespread heterozygote selective advantage leading to equilibria such as (5). Second, some variation could be accounted for by our observing selectively guided processes in which a superior gene is in the course of replacing an inferior gene. Other sorts of selective mechanisms not discussed above (for example, frequency-dependent selection) are also possible.

For most such selective explanations the problem arises of a large genetic load, either a segregational load or a substitutional load or a combination of the two. While the calculations of genetic load referred to above have been subject to a considerable reevaluation and modification, the load problem was perhaps the main argument (Kimura, 1968) for the introduction of the neutral theory. If the great majority of polymorphisms within populations are not caused by selective processes but rather by purely stochastic processes on selectively neutral genes, then presumably no load arises no matter how much genetic variation occurs within a population. The proponents of the neutral theory went even further than this. The second experimental advance of the mid-1960's, namely, protein sequencing, enables us to determine in fine detail the extent of genetic differentiation between different species and thus to estimate the number of evolutionary events (i.e., gene replacement processes) which have made up the course of evolution. Formal calculations of the substitutional load, given the number of replacements which are estimated to have taken place in the course of evolution, led to the claim that these substitutions could not all have been directed by natural selection, because an unbearably high substitutional load would have been incurred. The non-Darwinian theory claims, more precisely, that whereas there can be no doubt that gene replacement processes which have accounted for such obvious evolutionary phenomena as the development of the eye, the brain, and so on, have been directed by natural selection, there has also been a "background" process of "noise" evolution which in terms of gene frequency changes may well account for the great majority of substitutional processes, and which has had no relevance at all to the "improvement" of species and populations.

Thus the non-Darwinian theory has been supported by both static and evolutionary arguments. It follows that tests of the theory may be made by using "static" data and also by using "evolutionary" data. No single test on its own can then give a final verdict on the status of the theory. Such a verdict will be given, rather, by perhaps a rather subjective amalgamation of the results of these tests, which we now proceed to describe in some detail.

MATHEMATICS OF THE NEUTRAL THEORY: GENE FREQUENCY ANALYSIS

We consider in this section one of the ways in which the current gene frequencies in a population can be used to test the neutral theory.

The data one has at hand concerns a given gene locus in a given population, and gives the vector of frequencies of the various gene types observed in a sample of genes whose size is considerably less than that of the population. Our first aim must be to set up a model which reasonably describes the genetic evolution of a population under the assumption of neutrality. (Note that the model assumed will no doubt be to some extent arbitrary: nevertheless the test which it defines often applies for a wide range of neutral models.) From the predictions of this model we are then required to form a test of the neutral theory, based on gene frequency data. To do this, we must find the probability distribution of the data vector under the (null) hypothesis that the neutrality theory is true. Given this distribution, we can then use standard statistical methods to test whether or not the observed data conforms sufficiently closely to that expected under the neutral theory; if not, we reject the neutral theory and claim instead that evidence exists for selective differentials between the gene types. Second, our theory must be data-oriented; that is, it must apply to the form of data currently collected by geneticists. Such data is sample data rather than population data, since normally the geneticist does not have the resources to examine the entire population. Third, we must assume that the stochastic process governing gene frequencies in the population has reached stationarity. This assumption is rather severe and may not in fact apply for many real populations. For calculations on the rate of approach to stationarity, and two ways of overcoming problems of non-stationarity, see Ewens and Kirby (1975). Despite this problem, we assume from now on that stationarity has been reached.

The model we use derives from equation (11) and the assumption noted above concerning the novelty of new mutant types. It also assumes selective neutrality; in other words, we are developing the null hypothesis theory and distributions. Suppose in any generation there exist k gene types, with respective numbers $N_1, \ldots, N_k (\Sigma\, N_i = 2N)$. Then under the model we assume, the probability that in the next generation these gene types will occur with respective numbers $N_1', \ldots, N_k'$, and that further there are M mutant genes, is

$$Pr\{N_1' \cdots N_k', M\}$$

$$= \frac{(2N)!}{M!\,(N_1')! \cdots (N_k')!}\, u^M \prod_{i=1}^{k} \left\{ \frac{N_i}{2N}(i-u) \right\}^{N_i'}. \tag{16}$$

The model (16) could be called the generalized Wright-Fisher model of selectively neutral genes with novel mutations.

We have remarked that a sample of genes is supposed taken from a population obeying the model (16). This sample (of $2n$ genes, where we suppose $n \ll N$) will yield a data vector of the form $\{2n; k; n_1, \ldots, n_k\}$, where the notation means that in the sample of $2n$ genes, k different gene types were observed, and if these types are arbitrarily labelled $1, 2, \ldots, k$, that n_1 genes were observed of type $1, \ldots, n_k$ genes were observed of type k. Our problem now is to find the distribution of the vector $\{2n; k; n_1, \ldots, n_k\}$, assuming the model (16) and that stationarity has been reached. Once this has been done, the corresponding inductive question, namely how to use this distribution to test the hypothesis that the model (16) is correct (i.e., that selective neutrality is the case), can be attacked.

The derivation of the vector $\{2n; k; n_1, \ldots, n_k\}$ was obtained heuristically by Ewens (1972) and formally by Karlin and McGregor (1972). More specifically, Ewens used diffusion theory approximations, together with a conjecture which we discuss more precisely in the Appendix, to arrive at a conjectural distribution for this vector, and Karlin and McGregor then showed by a rather complex inductive argument that the distribution conjectured was in fact correct. Since this derivation provides an interesting use of diffusion theory, we outline in an Appendix firstly the way in which diffusion theory may be used to analyse problems in mathematical genetics, and secondly the application of this theory to the present problem. In order not to interrupt continuity, we proceed immediately to summarize the results reached in the Appendix.

Define the polynomial $S(\theta)$ by the equation

$$\begin{aligned} S(\theta) &= \theta(\theta+1)(\theta+2)\cdots(\theta+2n-1) \\ &= S_{2n}^{1}\theta + S_{2n}^{2}\theta^{2} + \cdots + S_{2n}^{2n}\theta^{2n}. \end{aligned} \tag{17}$$

Here the coefficients in the expansion of $S(\theta)$ are absolute values of Stirling numbers of the first kind. Then from equation (A12), the probability distribution of k is

$$Pr\{k\} = S_{2n}^{k}\,\theta^{k}/S(\theta),\; k = 1, 2, \ldots, 2n. \tag{18}$$

Suppose that k different gene types have been observed and are arbitrarily labelled types 1, 2, . . . , k. Let n_i be the number of genes of type i. Then the joint probability distribution of k, $n_1, \ldots, n_k$ is

$$Pr\{k, n_1, \ldots, n_k\} = \frac{(2n)!\theta^k}{k!S(\theta)n_1n_2 \cdots n_k}. \tag{19}$$

From (18) and (19) it follows immediately that the conditional distribution of $n_1, \ldots, n_k$, given k, is

$$Pr\{n_1, \ldots, n_k|k\} = \frac{(2n)!}{k!S_{2n}^k n_1n_2 \cdots n_k}. \tag{20}$$

Equations (18)–(20) are of fundamental importance in testing the neutral theory. Note that in all three equations the unknown parameters N and u enter only through the parameter θ. The value of θ is, of course, unknown and is subject to wide variation from one case to another. However, the distribution of $n_1, \ldots, n_k$, given k, is independent of θ, being given explicitly by (20). This fact allows us to use the data vector $\{k; n_1, \ldots, n_k\}$ as the basis for a test of neutrality: we simply check whether the values of $n_1, \ldots, n_k$, given k, conform sufficiently closely to the distribution (20).

To carry out such a test, it is necessary to construct some test statistic and to derive the neutral theory distribution of this test statistic from equation (20). The choice of such a statistic is to some extent arbitrary: one possibility is to use as a starting point the "information statistic"

$$I = -\sum_{i=1}^{k} x_i \log x_i, \tag{21}$$

where $x_i = n_i/2n$. In principle the neutral theory distribution of I can be constructed from (20), and thus the significance of any observed value of I obtained. In practice, technical problems arise and instead a test statistic having approximately the F distribution under the hypothesis of selective neutrality is used. Details are given in Ewens (1972).

Before applying this test, we should check whether the distribution (20) depends critically on the model (16) which we have used. The reason for this is that we have no real knowledge of what model nature "obeys", even under neutrality. Several models other than (16), in particular birth-and-death models, could quite possibly describe real-world behavior better than does (16). Fortunately it turns out that the result (20) is model-free: so long as the stochastic model describing the way in which the genetic composition of the daughter generation derives from that of the parent generation incorporates fundamentally the concept of selective neutrality, equation (20) holds true. This result is derived by Kirby and Ewens (1976) and gives us considerable confidence that equation (20) can be used uniformly in our hypothesis testing procedures.

Despite this, it is probably premature to apply the testing procedure derived from (20) to present data. This is because the technique of electrophoresis, which has been used to provide almost all of present-day data, does not always differentiate two different gene types (it will do so only if certain electrical charge properties of these types are different). The model we have just described assumes that the complete identification of different gene types is possible, and it may be shown that a bias is introduced by applying it to electrophoretic data.

Two strategies are now possible. One is to continue to develop the above theory in the hope that technical advances will soon lead to the construction of an apparatus having complete power of differentiation; considerable progress has been made in this direction. The second approach is to construct a more complex mathematical model which hopefully describes more exactly the behavior of electrophoretically derived gene frequencies under selective neutrality. Although one or two elegant results in this direction exist (see Ohta and Kimura, 1973), it is possible that the mathematical complexity of these models will be such that the first strategy will be more useful. In any event, electrophoresis is merely a phase in experimental development and it seems more appropriate not to concentrate on theoretical results whose appropriateness is temporary.

MATHEMATICS OF THE NEUTRAL THEORY: EVOLUTIONARY ANALYSIS

The neutral theory was inspired in part by the nature and extent of interspecific genetic variation and in this section we indicate how this occurred and the way in which mathematical tests can be constructed based on interspecific data.

In the model (16) the mean number of new mutant types to arise per generation is $2Nu$, and if u is extremely small, the probability that any particular one of these types reaches a frequency of unity can be taken, from the argument preceding equation (15), to be $(2N)^{-1}$. Thus the mean number of new types to arise in any generation and eventually reach fixation is $(2Nu)(2N)^{-1} = u$. Equivalently we can see that the mean time, measured in generations, between such evolutionary events is u^{-1}. Note that this value is independent of N and, since we assume selective neutrality, is of course also free of the values of any selective parameters.

On the other hand, the corresponding calculation for a selective locus does not give so simple an answer. Using equation (15), the mean time between successive evolutionary events controlled by selection is of the form $[4Nv(w_{12} - 1)]^{-1}$, where v is the mutation rate to selectively advantageous types. Clearly this will vary greatly from one case to another: the rate for neutral genes, on the other hand, is relatively stable. We thus have different predictions under the two models. Which is closer to the observed facts of evolution? Neutralists claim that on the molecular level evolution has been remarkably uniform, and this was indeed one of the major arguments which led to the putting forward of the neutral theory. In order to arrive at an objective assessment a formal testing procedure is, however, required; this has recently been supplied by Langley and Fitch (1974). Under Langley and Fitch's procedure one first estimates the historical evolutionary pattern which led to the present-day observations on the genetic make-up of contemporary species. This pattern is in fact an evolutionary tree which provides "observed" numbers of evolutionary events in its various branches. These "observed" values can be compared to those expected under the uniform (neutral) prediction, using a chi-square test. When this is done it turns out that significant departures from uniformity are found. This conclusion supports the selective theory. Neutralists nevertheless claim that despite this the observed rates are still sufficiently uniform, given the number of unknown parameters and assumptions implicit in the model, that neutrality cannot be rejected out of hand. The argument still continues and no doubt further theoretical developments will be necessary before it is finally settled.

CONCLUSION

The neutralist-selectionist controversy still continues. Before it is finally settled a number of theoretical results not yet developed will be required. The above brief discussion outlines such theoretical results as are presently known. The interested reader should consult the references given below for further reading on this topic.

Appendix

In this Appendix we discuss the way in which diffusion theory can be used to theoretical population genetics, and thus outline the derivation of equations (18) and (19). Our approach will be non-rigorous and heuristic.

Suppose x is a Markovian random variable in [0,1], and that at discrete time points $\delta t, 2\delta t, 3\delta t, \ldots, x$ can change its value according to some finite Markov chain probability structure. Suppose that the transition probabilities of x are such that

$$E\left[x(t+\delta t) - x(t) \mid x(t) = x\right] = m(x)\delta t + O(\delta t)^2,$$

$$\text{Var}\left[x(t+\delta t) - x(t) \mid x(t) = x\right] = v(x)\delta t + O(\delta t)^2,$$

$$E\left[\{x(t+\delta t) - x(t)\}^i \mid x(t) = x\right] = o(\delta t), i \geqslant 3. \qquad \text{(A1)}$$

If we write $x(t+\delta t) - x(t)$ as δx, and if we write $p(x;\, t)$ for the probability that the random variable assumes the value x at time t, we have from the theorem of total probability

$$p(x; t + \delta t) = \Sigma_{\delta x} p(x - \delta x; t) g(\delta x; x - \delta x), \qquad \text{(A2)}$$

where $g(\delta x; x - \delta x)$ is the probability of a change in the value of the random variable from $x - \delta x$ to x in one step. If we expand the right hand side in (A2) as a Taylor series as (in slightly abbreviated notation)

$$\begin{aligned} p(x - \delta x) g(x - \delta x) &= p(x) g(x) - (\delta x)[p(x) g(x)]' \\ &\quad + \tfrac{1}{2}(\delta x)^2 [p(x) g(x)]'' + o(\delta x)^2 \end{aligned}$$

and then perform the summation, we obtain, using (A1),

$$\begin{aligned} p(x; t + \delta t) = p(x; t) + \delta t \biggl[&- \frac{\partial}{\partial x}\{m(x) p(x; t)\} \\ &+ \tfrac{1}{2} \frac{\partial^2}{\partial x^2}\{v(x) p(x; t)\} \biggr] + O(\delta t)^2. \qquad \text{(A3)} \end{aligned}$$

Suppose now we let $\delta t \to 0$, so that the Markov process becomes a continuous time continuous state diffusion process. Dividing by δt in (A3) and writing now $f(x; t)$ as the density function of x at time t, we obtain

$$\frac{\partial}{\partial t} f(x; t) = -\frac{\partial}{\partial x}\{m(x) f(x; t)\} + \tfrac{1}{2} \frac{\partial^2}{\partial x^2}\{v(x) f(x; t)\}. \qquad \text{(A4)}$$

Equation (A4) is the celebrated Fokker-Planck, or forward Kolmogorov, diffusion equation. It is of the greatest importance in theoretical population genetics since its complete solution, or at any rate various important properties of its solution, can be found in a wide variety of genetical models.

To give an example of how a discrete process in genetics can be approximated by a diffusion model described by (A4), we take the process in which we are currently interested, namely that described by (16). It is clear that the number N_i of genes of the ith type is a Markovian random variable. Further it is evident that if we introduce the scaled random variables $x_i = N_i/2N$, $x_i' = N_i'/2N$, then

$$E(x_i' - x_i) = -ux_i,$$

$$\mathrm{Var}\,(x_i' - x_i) = x_i(1 - x_i)/(2N) + O(uN^{-1}),$$

with higher moments of order N^{-2} or less. To fit this model into the framework assumed by (A1), it is necessary to assume u is of order N^{-1} and that time is measured in units of $2N$ generations (i.e.,$\delta t = (2N)^{-1}$). If we put $\theta = 4Nu$, the equation $m(x)\delta t = -ux$ then yields $m(x) = -\frac{1}{2}\theta x$, and similarly $v(x) = x(1 - x)$. It follows that if x_i is now treated as a continuous random variable, its density function at time t (= $2Nt$ generations) satisfies, in the diffusion limit, the equation

$$\frac{\partial}{\partial t} f(x; t) = -\frac{\partial}{\partial x}\left\{-\tfrac{1}{2}\theta x f(x; t)\right\}$$
$$+ \tfrac{1}{2} - \frac{\partial^2}{\partial x^2}\{x(1 - x)f(x; t)\}. \quad \text{(A5)}$$

Equation (A5) can be solved exactly by separation of variables, the solution involving an infinite series of hypergeometric functions with appropriate eigenvalue weightings. Details, (making various notational changes and considering $1 - x$ rather than x), are given by Crow and Kimura (1970, page 394).

In the present context we are interested in the following question: what is the mean time that the frequency x takes values in the interval (x_1, x_2), before its eventual (and certain) absorption at the value of 0? In principle this can be evaluated by considering $\int_{x1}^{x2} \int_0^\infty f(x, t)dt\, dx$. However a more direct approach to finding such mean time properties is available. We do not pursue the details (for these, see Ewens (1969, pp. 53–55)), and note that in our case that if p is the initial value of x, and $p \leqslant x_1 < x_2 \leqslant 1$, the required mean time is

$$2\int_{x_2}^{x_1} x^{-1}(1-x)^{\theta-1}dx \text{ generations.} \tag{A6}$$

We now use equation (A6) in the following way. In our case $p = (2N)^{-1}$, since each gene type is initially represented by a single (mutant) gene. Further, the mean number of new types to appear by mutation in each generation is $2Nu$. An elementary ergodic argument then shows that the mean number of different types present in any generation whose frequency is in (x_1, x_2) is

$$\theta\int_{x_1}^{x_2} x^{-1}(1-x)^{\theta-1}dx, \qquad (2N)^{-1} \leqslant x_1 < x_2 \leqslant 1. \tag{A7}$$

This expression is fundamental to the heuristic development of the sampling theory of the data vector $\{2n; k; n_1, \ldots, n_k\}$ defined in the main text. We suppose that $n \ll N$ so that, although sampling is without replacement, binomial formulae can be used to a sufficiently close approximation.

We suppose a sample of j genes is taken from the population. The probability that these are all of the same allelic type is, from (A7),

$$\theta\int_0^1 x^j\left[x^{-1}(1-x)^{\theta-1}\right]dx$$

$$= \theta(j-1)!/\left[\theta(\theta+1)\cdots(\theta+j-1)\right]. \tag{A8}$$

It follows from (A8) that the probability that a sample of $j+1$ genes drawn in a sample are all of the same type, given that the first j genes are all the same, is

$$Pr\{(j+1)\text{th gene not new}|\text{first } j \text{ identical}\} = j/(\theta+j). \tag{A9}$$

Now consider the probability that, irrespective of the first j genes sampled, the $(j+1)$th gene will be of an allelic type not previously observed. Again using (A7), this probability is

$$Pr\{(j+1)\text{th gene new}\} = \theta\int_0^1 (1-x)^j x\left[x^{-1}(1-x)^{\theta-1}\right]dx$$
$$= \theta/(\theta+j).$$

It follows immediately that

$$Pr\{(j+1)\text{th gene not new}\} = j/(\theta+j). \tag{A10}$$

Comparison of (A9) and (A10) yields the initially surprising conclusion that the probability that a gene of a new type is seen on any draw is not changed if we are given the information that all genes previously drawn were of identical type. The conjecture we now make, based on this observation, is that (A10) is true *whatever* information is given about the nature of the first j genes drawn. Surprisingly, this conjecture is true, although we are not yet in a position to prove this here.

let $q_{j,i}$ be the probability that a sample of j genes yields i different types. Then under the above conjecture

$$q_{j+1,i} = q_{j,i}\{j/(\theta+j)\} + q_{j,i-1}\{\theta/(\theta+j)\}. \tag{A11}$$

The recurrence relation (A11), subject to obvious boundary conditions, can be solved readily. Since we are interested in a sample of size $2n$, we are particularly interested in the value $q_{2n,k}$. We find

$$q_{2n,k} = Pr\{k\} = S_{2n}^k\theta^k/S(\theta), \tag{A12}$$

where S_{2n}^k and $S(\theta)$ are defined in equation (17).

It is of course far more complex to find the distribution of the entire vector $\{2n; k; n_1, \ldots, n_k\}$, but this can be done by an extension of the argument given (continuing to assume that the conjecture made after equation (A10) is true). The probability distribution which results is (19). (The argument, which is long but straightforward, is given in Appendix 2 of Ewens, 1972.)

We have thus arrived at (19) on a heuristic basis. The argument is now completed formally by showing by induction that (19) is in fact true. The complete inductive proof is given by Karlin and

McGregor (1972) and is too long to be reproduced here. The argument used is that, with probability $1 - O(N^{-1})$, the sample of $2n$ genes must emanate from either $2n$ or $2n - 1$ "parent" genes. A relationship is then derived between the probability (19) for a sample of $2n - 1$ genes and one for $2n$ genes. This leads to the inductive proof since (19) is clearly true for $2n = 1$.

REFERENCES

1. A. C. Allison, "Aspects of polymorphism in man," *Cold Spring Harbor Symp. Quant. Biol.*, **20** (1955), 229–235.
2. F. J. Ayala and J. R. Powell, "Allozymes as diagnostic characters of sibling species of Drosophila," *Proc. Nat. Acad. Sci. U.S.A.*, **69** (1972), 1094–1096.
3. L. L. Cavalli-Sforza and W. F. Bodmer, *The Genetics of Human Populations*, Freeman, San Francisco, 1971.
4. J. F. Crow and M. Kimura, *An Introduction to Population Genetics Theory*, Harper and Row, New York, 1970.
5. Th. Dobzhansky, *Genetics of the Evolutionary Process*, Columbia University Press, New York, 1970.
6. W. J. Ewens, "Random sampling and the rate of gene replacement," *Evolution*, **21** (1967), 657–663.
7. ———, *Population Genetics*, Methuen, London, 1969.
8. ———, "The sampling theory of selectively neutral alleles," *Theoret. Pop. Biol.*, **3** (1972), 87–112.
9. ———, "Mathematical and statistical problems arising in the non-Darwinian theory," *Lectures on Mathematics in the Life Sciences*, **7** (1974), 25–42.
10. W. J. Ewens and K. Kirby, "The eigenvalues of the neutral alleles process," *Theoret. Pop. Biol.*, **7** (1975), 212–220.
11. R. A. Fisher, "On the dominance ratio," *Proc. Roy. Soc. Edinburgh*, **42** (1922), 321–341.
12. ———, *The Genetical Theory of Natural Selection*, Clarendon Press, Oxford, 1930.
13. J. Gillespie and K. Kojima, "The degree of polymorphism in enzymes involved in energy production compared to that in nonspecific enzymes in two D. ananassae populations," *Proc. Nat. Acad. Sci. U.S.A.*, **61** (1968), 582–585.
14. J. B. S. Haldane, *The Causes of Evolution*, Longmans Green, London, 1932.
15. G. H. Hardy, "Mendelian proportions in a mixed population," *Science*, **28** (1908), 49–50.
16. H. Harris, "Enzyme polymorphisms in man," *Proc. Roy. Soc. Ser. B.*, **164** (1966), 298–310.

17. ———, "Annotation: polymorphism and protein evolution: the neutral mutation-random drift hypothesis," *J. Med. Genet.*, **8** (1971), 444–452.

18. S. Karlin and J. McGregor, "Addendum to a paper of W. J. Ewens," *Theoret. Pop. Biol.*, **3** (1972), 113–116.

19. M. Kimura, "Evolutionary rate at the molecular level," *Nature*, **217** (1968), 624–626.

20. M. Kimura and J. F. Crow, "The number of alleles that can be maintained in a finite population," *Genetics*, **49** (1964), 725–738.

21. M. Kimura and T. Ohta, *Theoretical Aspects of Population Genetics*, Princeton University Press, Princeton, 1971.

22. ———, *Population Genetics, Molecular Biometry and Evolution, VI Berkeley Symposium on Probability and Statistics*, ed. L. Le Cam, J. Neyman, and E. L. Scott, U. of California, Berkeley, 1972, 43–68.

23. J. L. King and T. H. Jukes, "Non-Darwinian evolution: random fixation of selectivity neutral mutations," *Science*, **164** (1969), 788–798.

24. K. Kirby and W. J. Ewens, 1976, (to appear).

25. K. Kojima, J. Gillespie, and Y. N. Tobari, "A profile of Drosophila species enzymes assayed by electrophoresis I. Number of alleles, heterozygosities and linkage disequilibrium in glucose-metabolizing systems and some other enzymes," *Biochem. Genet.*, **4** (1970), 627–637.

26. C. H. Langley and W. M. Fitch, "An examination of the constancy of the rate of molecular evolution," *J. Mol. Evol.*, **3** (1974), 161–177.

27. R. C. Lewontin and J. L. Hubby, "A molecular approach to the study of genic heterozygosity in natural populations. II. Amount of variation and degree of heterozygosity in natural populations of Drosophila pseudoobscura," *Genetics*, **54** (1966), 595–609.

28. R. C. Lewontin, "Testing the theory of natural selection," *Nature*, **236** (1972b), 181–182.

29. ———, *The Genetic Basis of Evolutionary Change*, Columbia University Press, 1974.

30. T. Ohta and M. Kimura,"A new model for estimating the number of electrophoretically detectable alleles in a finite population," *Proc. 13th Internat'l Cong. of Genet. Abstracts, Genetics*, **74** (1973), No. 2, pt. 2, S201.

31. J. A. Sved, T. F. Reed, and W. F. Bodmer, "The number of balanced polymorphisms that can be maintained in a natural population," *Genetics*, **55** (1967), 469–481.

32. G. A. Watterson, "The sampling theory of selectively neutral alleles," *Advances in Appl. Probability*, **6** (1974), 463–488.

33. W. Weinberg, "On the laws of heredity in man (German)," *Z. fuer induktive abstammungs und vererbungslehre*, **1** (1908), 377–392.

34. S. Wright, "Evolution in mendelian populations," *Genetics*, **16** (1931), 97–159.

THEORETICAL ASPECTS OF MULTI-LOCUS SELECTION BALANCE I*

Samuel Karlin

The last 20 years have witnessed extensive theoretical studies of complex genetic systems integrating the interactive effects of several agents and relations among loci.† The investigation of general multi-locus selection regimes remains an important open problem of population genetics. Although recognized early and widely by experimental and theoretical population geneticists, that fitness generally involves the interactive effects of many loci coupled to a myriad of exogenous and endogenous environmental factors, this point has been dramatized in the paper of Franklin and Lewontin [1970], "Is the gene the unit of selection?", and further underscored in the recent book by Lewontin [1974].

A number of classes of two-locus selection models were studied by Wright [1952], Kimura [1956], Lewontin and Kojima [1960], Bodmer and Felsenstein [1967], Karlin and Feldman [1970b] and

*Supported in part by NIH Grant GM10452-13 and NSF Grant MPS71-02905A03

†We refer to Karlin [1972] for an elementary review of some classical mathematical models of population genetics, representative references, and most importantly, relevant terminology.

others. Several findings on two-locus theory with a *general* viability regimes are elaborated in Karlin [1975] and numerical studies set forth in Karlin and Carmelli [1975b]. An important contribution to the study of the multi-locus multiplicative nonepistatic selection model is the work of Roux [1974]. Analytic studies of other two- and three-locus viability systems are found in Feldman, Franklin and Thomson [1974], Feldman *et al.* [1975], Christiansen and Feldman [1975], Strobeck [1976], among others. For an increasing set of selection schemes it has been possible to develop a representation of the chromosome frequency transformation in canonical algebraic forms entailing sums of Kronecker products. The first use of this technique appears in Roux [1974]. Karlin [1977a] and Karlin and Lieberman [1977] have refined and extended the formulation and methodology permitting the ascertainment of much of the equilibrium structure for a hierarchy of nonepistatic multiplicative and additive viability schemes. These approaches lead to various characterizations of stable equilibrium configurations for certain combinations of multi-locus phenotypic selection expressions (see Sections 6 and 7). In terms of appropriate representation formulas we are also able to study viability schemes which have mixed additive and multiplicative effects among a series of loci, various grades of non-epistatic to epistatic selection patterns, and selection regimes where the recombination parameters differ in the two sexes. A number of other models involving two or more loci have been examined numerically and/or by simulation methods. Although highly nonlinear and complex, the mathematical treatment of these models, buttressed by statistical and numerical approaches, is significantly tractable.

The accumulation of experimental and field observation on multi-locus traits (especially systems of enzyme polymorphisms) calls for more discriminating analyses and interpretation. In this respect it is necessary to expand the theory seeking to establish some principles and robust conclusions that help to discern the nature of the selection influence and the role of the recombination mechanism. Moreover, a carefully designed numerical study of various classes of selection schemes entailing different recombination assumptions can be evaluated more meaningfully where parallel theoretical insights serve as a control.

In this manuscript we report a series of results relevant to the dynamic and equilibrium behavior of gamete frequency configurations associated with various classes of multi-locus multi-allele viability patterns and recombination schemes. Proofs of the results presented here as well as further biological elaborations and speculations deriving from our analytic findings will mostly appear in Karlin [1979].

We indicate now the organization of this paper. The one-locus multi-allele model is formulated in Section 1. Section 2 reviews a number of known facts in this matter. Sections 3 and 4 delineate a hierarchy of refinements and elaborations of the theory vital to our later treatment of multi-locus systems. Relevant facets of the two-locus selection model are covered in Section 5. Sections 6 and 7 highlight a series of results pertaining to the nature of the equilibrium gene frequency arrays associated with certain classes of multi-locus viability regimes in concert with different stipulations on sets of recombination rates operating between loci. Some concepts in measuring association of loci concomitant to disequilibrium indices of gene frequency are implicated.

Discussions in Sections 1–8 provide a glimpse of an assortment of results, various approaches, and qualitative and quantitative interpretations relevant to the theory of complex genetic systems. We have left out several topics pertinent for a more complete perspective and understanding of multi-locus selection theory. Here is a partial list and some relevant references:

(i) Theory of modifier genes (e.g., Nei [1967], Feldman [1972], Karlin and McGregor [1972b, 1974], and Feldman and Krakauer [1976]).
(ii) The advantages and disadvantages of recombination with respect to the incorporation of favorable mutations, e.g., Eshel and Feldman [1970], J. Maynard Smith [1971], Karlin [1973].
(iii) Multi-locus haploid models, e.g., Feldman [1971].
(iv) The effect of other mating systems in multi-locus contexts as selfing, assortative, and incompatibility mating patterns and sexual selection schemes (e.g., see Karlin [1968], Karlin and Farkash [1978]).

(v) Conditions for two-locus mutation selection balance (Karlin and McGregor [1971c]), and more generally the interaction of mutation, migration, and selection forces in multi-locus systems (e.g., Li and Nei [1974], Christiansen and Feldman [1974]).

(vi) The interactive effects of natural and artificial selection forces on multi-locus traits (e.g., Karlin and Carmelli [1975a]).

(vii) Two- and multi-locus systems in the presence of varying vis-à-vis constant environments as expressed through temporal and spatial random fluctuations of genotypic selection intensities (e.g., see Karlin and Lieberman [1974], Gillespie and Langley [1976]).

(viii) Finite population size (in multi-locus contexts) and sampling fluctuation effects, associated overdominance, etc. (e.g., Ohta and Kimura [1969], Sved [1968], Hill [1974, 1976], and Karlin and McGregor [1968]).

(ix) The connections of quantitative and polygenic inheritance to multi-locus phenomena (e.g., Lande [1976], Karlin [1978]).

1. ONE-LOCUS MULTI-ALLELE SELECTION MODELS: FORMULATION

The formulation of the first models in Sections 1 and 2 traces back at least 30 years. Substantial progress in discerning the nature of selection balance has been made for a number of multi-allele one-locus systems. Notable contributions here include Kingman [1961a], [1961b], Mandel [1959], Kimura [1958] and others. See Crow and Kimura [1970] for further references and background material. The discussion of the general model in paragraph III below may be partly new.

Consider a bisexual (two separated sexes) population characterized by r possible alleles (= gamete types) $A_1, A_2, \ldots, A_r$ at a given locus and, of course, $r(r+1)/2$ genotypes $\{A_iA_j\}$. Suppose at the start of a generation the frequency vector of the r allele array in the female population is $(p_1, p_2, \ldots, p_r)$ and $(q_1, q_2, \ldots, q_r)$ for the male population. Let the evolution of the population be subject to the following influences: (i) random union of gametes in discrete non-overlapping generations with Mendelian segregation and 1:1 sex ratio*; (ii) differential viability selection. The viability fitness matrix for females is $F = \|f_{ij}\|_{i,j=1}^{r}$, where f_{ij} measures the relative average number of the A_iA_j genotype that survive from infancy to maturity. The viability fitness matrix for males is designated by $M = \|m_{ij}\|$. To avoid unimportant technical details we stipulate throughout (unless stated otherwise) that $f_{ij} > 0$ and $m_{ij} > 0$ for all i, j. It is clear from their meaning that F and M are symmetric matrices.

I. The classical multi-allele selection model for an autosomal character. From the stipulation of random union of gametes a typical genotype A_iA_j is created with frequency $p_iq_j + q_ip_j$, $i \neq j$, of which half are females (because of the assumption of equal sex ratio). As a consequence of viability selection, the relative frequency of A_iA_j among mature females is $[p_iq_j + q_ip_j]f_{ij}$ and the proportion of A_i gametes among these contributed by Mendelian segregation is $\frac{1}{2}[p_iq_j + q_ip_j]f_{ij}$. It follows that the frequency p_i' of the A_i gamete in the next generation is proportional to $\frac{1}{2}\sum_{j=1}^{r}(p_iq_j + q_ip_j)f_{ij}$ or after normalization

$$p_i' = \frac{\frac{1}{2}\left[p_i\left(\sum_{j=1}^{r} f_{ij}q_j\right) + q_i\left(\sum_{j=1}^{r} f_{ij}p_j\right)\right]}{\sum_{i,j=1}^{r} f_{ij}p_iq_j}, \quad i = 1, 2, \ldots, r. \tag{1.1}$$

Similarly,

$$q_i' = \frac{\frac{1}{2}\left[p_i\sum_{j=1}^{r} m_{ij}q_j + q_i\sum_{j=1}^{r} m_{ij}p_j\right]}{\sum_{i,j=1}^{r} m_{ij}p_iq_j}, \quad i = 1, 2, \ldots, r. \tag{1.2}$$

*For the viability model of this paragraph, random mating of zygote (genotype) individuals and random union of gametes lead to equivalent formulations; cf. paragraph III later.

It will be convenient to use compact vector notation. *The Schur product of the vectors* $a = (a_1, \ldots, a_r)$ *and* $b = (b_1, \ldots, b_r)$ *is symbolized by* $a \circ b = (a_1 b_1, a_2 b_2, \ldots, a_r b_r)$. *The inner product of two vectors is denoted by* $(a, b) = \Sigma_{i=1}^{r} a_i b_i$. In this notation the recursion relations of (1.1) and (1.2) take the form

$$p' = \tfrac{1}{2}\,\frac{p \circ Fq + q \circ Fp}{(p, Fq)}, \quad q' = \tfrac{1}{2}\,\frac{q \circ Mp + p \circ Mq}{(p, Mq)}. \tag{1.3}$$

The expression (1.3) displays a non-linear transformation T of $2r$ variables, ($(2r\text{-}2)$ independent ones, since $\Sigma p_i = \Sigma q_i = 1$) which we abbreviate in the form

$$\{p', q'\} = T\{p, q\}.$$

In the case where $M = F$ (such that selection operates on the male and female genotypes in the same way) it is manifest from (1.3) that $p' = q'$ and thereafter the transformation T can be reduced to the single set

$$p' = \frac{p \circ Mp}{(p, Mp)}, \tag{1.4}$$

where the sexes are not distinguished. We discuss the model of (1.4) in some detail in Sections 2 and 3.

II. A multi-allele selection model for a sex linked character. This model was first formulated by Haldane; consult also Cannings [1967], [1969b]. Consider a character determined by a locus on the sex chromosome with r possible alleles as before. Suppose the female sex is the homogametic one, namely carries the XX chromosomes, while the male is heterogametic, carrying the XY chromosomes, and only an X contributes an allele.

Accordingly, the female genotypes in the population are A_iA_j, $i, j = 1, \ldots, r$, as previously, but the male genotypes take the form A_iY where the Y chromosome carries no complement of the gene.

The viability coefficients corresponding to females are summarized by the matrix $F = \|f_{ij}\|$ while for males by the vector

$m = (m_1, m_2, \ldots, m_r)$. Thus, m_i measures the relative fitness of the male genotype $A_i Y$ and f_{jk} of the female genotype $A_j A_k$. Subject to random union of gametes and viability selection, with Mendelian segregation we obtain the transformation law

$$p_i' = \frac{\frac{1}{2}\left[\left(p_i \sum_{j=1}^{r} f_{ij} q_j\right) + q_i\left(\sum_{j=1}^{r} f_{ij} p_j\right)\right]}{\sum_{i,j=1}^{r} p_i f_{ij} q_j}, \quad q_i' = \frac{m_i p_i}{\sum_{i=1}^{r} m_i p_i}.$$

In our compact vector notation this reads as

$$p' = \frac{\frac{1}{2}\left[p \circ Fq + q \circ Fp\right]}{(p, Fq)}, q' = \frac{m \circ p}{(m, p)}. \tag{1.5}$$

III. A general selection model. The remainder of this section is not pertinent to the discussion of the later sections and is only inserted to provide wider perspective on the nature of natural selection forces acting on populations. It is useful to formulate a general selection model that incorporates the three common forms of natural selection pressures: differential viabilities, differential fertilities and segregation distortion. Viability coefficients f_{ij} and m_{ij} for the male and female $A_i A_j$ genotype, $i, j = 1, 2, \ldots,$ respectively, reflect the relative abilities for the alternative genotypes to survive from infancy to maturity. The various fertility intensities indicate the differential mean number of infant offspring produced by a mating type. More specifically, associated with each mating $(A_i A_j) \times (A_k A_l)$ is the measure $\phi(i, j;\ k, l)$ of the mean number of progeny of the mating. In the most general formulation we can have $\phi(i, j;\ k, l) \neq \phi(k, l;\ i, j)$ so that the order of the parental genotypes (the dependence on which is the male and female parental genotypes) is important where reciprocal crosses may entail different mean fertility outcomes. Segregation distortion signifies that the proportions of the zygotic possibilities in the cross $(A_i A_j) \times (A_k A_l)$ may differ from an equally likely combination of gametes one contributed by each parent. Under Mendelian segregation the segregation output can be displayed in the form

$$(A_iA_j) \times (A_kA_l) \to \tfrac{1}{4}(A_iA_k) + \tfrac{1}{4}(A_iA_l) + \tfrac{1}{4}(A_jA_k) + \tfrac{1}{4}(A_jA_l) \tag{1.6}$$

(i.e., with chance $\frac{1}{4}$ the zygote A_iA_k is produced, with chance $\frac{1}{4}$ A_iA_l, etc.).

With segregation distortion, we can have the progeny ratios

$$A_iA_j \times A_kA_l \to \alpha(i, k; j, l)(A_iA_k) \\ + \alpha(i, l; j, k)(A_iA_l) + \alpha(j, k; i, l)(A_jA_k) + \alpha(j, l; i, k)(A_jA_l) \tag{1.7}$$

where the α's above add to 1, but they are not all equal. For the general selection model involving all three forms of selection forces listed, it is *not true* that *random mating* (= random formation of genotype crosses) *is equivalent* to *random union of gametes*. Conditions where this equivalence holds will be indicated shortly.

Let $\bar{P}_{ij}$ = frequency of A_iA_j (in this order) in the infant female population and $P_{ij} = \bar{P}_{ij} + \bar{P}_{ji}$ (= frequency of the unordered A_iA_j when $i \neq j$ and twice the frequency of A_iA_i if $i = j$). The quantities $\bar{Q}_{ij}$ and Q_{ij} are defined analogously with reference to the male genotype. Differential viability selection acts in conversion from infants to adults. Let $\tilde{P}_{ij}$ and $\tilde{Q}_{ij}$ denote the frequencies of the associated mature genotypes. In line with the development of (1.1), we obtain

$$\tilde{P}_{ij} = \frac{\bar{P}_{ij}f_{ij}}{\sum_{i,j} \bar{P}_{ij}f_{ij}}, \quad \tilde{Q}_{ij} = \frac{\bar{Q}_{ij}m_{ij}}{\sum_{i,j} \bar{Q}_{ij}m_{ij}}. \tag{1.8}$$

Random mating signifies that the cross $A_iA_j \times A_kA_l$ (the ordering is significant where the first genotype corresponds to the female involved and the second to the male, respectively) occurs with frequency $\tilde{P}_{ij}\tilde{Q}_{kl}$, the relative number of offspring being

$\phi(i, j; k, l)$. We allow general segregation distortion as in (1.7). The number of A_jA_i female infants (the ordering will signify, consistent with the preceding, that the first gamete A_j derives from the female parent and A_i is contributed by the male parent) is

$$P'_{ji} = \frac{1}{N}\left\{\sum_{\mu,\nu} \tilde{P}_{\mu j}\tilde{Q}_{\nu i}\phi(\mu, j; \nu, i)\alpha(j, i; \mu, \nu)\right.$$

$$+ \sum_{\mu,\nu} \tilde{P}_{\mu j}\tilde{Q}_{i\nu}\phi(\mu, j; i, \nu)\alpha(j, i; \mu, \nu)$$

$$+ \sum_{\mu,\nu} \tilde{P}_{j\mu}\tilde{Q}_{\nu i}\phi(j, \mu; \nu, i)\alpha(j, i; \mu, \nu)$$

$$\left.+ \sum_{\mu,\nu} \tilde{P}_{j\mu}\tilde{Q}_{i\nu}\phi(j, \mu; i, \nu)\alpha(j, i; \mu, \nu)\right\} \tag{1.9}$$

where

$$N = 4 \sum_{i,j,\mu,\nu} \tilde{P}_{j\mu}\tilde{Q}_{i\nu}\phi(j, \mu; i, \nu)\alpha(j, i; \mu, \nu).$$

A similar formula applies to obtain the equation for Q'_{ji}.

We now concentrate on the special case where the fertility factors are the products.

$$\phi(i, j; k, l) = \psi(i, j)\theta(k, l) \tag{1.10}$$

with a fecundity factor $\psi(i, j)$ attributable to the female parent and $\theta(k, l)$ the fecundity factor associated with the male parent where the effects operate independently. We stipulate the natural symmetry identities $\psi(i, j) = \psi(j, i)$, $\theta(i, j) = \theta(j, i)$. With Mendelian segregation (the factorization $\alpha(i, j; k, l) = \alpha(i)\alpha(j)\alpha(k)\alpha(l)$ would suffice) the combined transformation laws (1.8) and (1.9) can be considerably reduced yielding

$$\bar{P}'_{ij} = \frac{1}{N^*}\sum_{\mu,\nu}\left[\bar{P}_{\mu i}\bar{Q}_{\nu j}B_1 + \bar{P}_{\mu i}\bar{Q}_{j\nu}B_2 + \bar{P}_{i\mu}\bar{Q}_{\nu j}B_3 + \bar{P}_{i\mu}\bar{Q}_{j\nu}B_4\right], \tag{1.11}$$

where

$$B_1 = f_{\mu i} m_{\nu j} \psi(\mu, i) \theta(\nu, j),\ B_2 = f_{\mu i} m_{j\nu} \psi(\mu, i) \theta(j, \nu),$$
$$B_3 = f_{i\mu} m_{\nu j} \psi(i, \mu) \theta(\nu, j),\ B_4 = f_{i\mu} m_{j\nu} \psi(i, \mu) \theta(j, \nu).$$

But $B_1 = B_2 = B_3 = B_4$ on account of the imposed symmetry identities and (1.11) becomes

$$\bar{P}'_{ij} = \frac{1}{N^*} \sum_{\mu, \nu} \left(\bar{P}_{\mu i} + \bar{P}_{i\mu}\right)\left(\bar{Q}_{\nu j} + \bar{Q}_{j\mu}\right) B_1$$
$$= \frac{1}{N^*} \left(\sum_{\mu} P_{\mu i} v_{\mu i}\right)\left(\sum_{\nu} Q_{\nu j} w_{\nu j}\right), \tag{1.12}$$

where

$$v_{\mu i} = f_{\mu i} \psi(\mu, i),\ w_{\nu j} = m_{\nu j} \theta(\nu, j),$$

and

$$N^* = \left(\sum_{\mu, i} P_{\mu i} v_{\mu i}\right)\left(\sum_{\nu, j} Q_{\nu j} w_{\nu j}\right).$$

Defining

$$p'_i = \frac{\sum_{\mu} P_{\mu i} v_{\mu i}}{\sum_{\mu, i} P_{\mu i} v_{\mu i}},\ q'_j = \frac{\sum_{\nu} Q_{\nu j} w_{\nu j}}{\sum_{\nu, j} Q_{\nu j} w_{\nu j}}, \tag{1.13}$$

we obtain

$$\bar{P}'_{ij} = p'_i q'_j$$

and

$$P'_{ij} = p'_i q'_j + p'_j q'_i \text{ (and therefore } = Q'_{ij}). \tag{1.14}$$

In terms of the variables (1.13) we readily compute p''_i for the succeeding generation to be

$$p''_i = \frac{\sum_{\mu} (p'_\mu q'_i + p'_i q'_\mu) v_{\mu i}}{\sum_{\mu, i} (p'_\mu q'_i + p'_i q'_\mu) v_{\mu i}}$$

$$= \frac{1}{2} \frac{\sum_{\mu} (p'_\mu q'_i + p'_i q'_\mu) v_{\mu i}}{\sum_{\mu, i} p'_\mu q'_i v_{\mu i}}. \tag{1.15}$$

The same transformation (with $w_{\nu j}$ instead of $v_{\mu i}$) connects q''_j to q'_j in the manner

$$q''_j = \frac{1}{2} \frac{\sum_{\nu} (p'_\nu q'_j + p'_j q'_\nu) w_{\nu j}}{\sum_{\nu, j} p'_\nu q'_j w_{\nu j}}. \tag{1.16}$$

Comparing (1.15) and (1.16) to (1.3) we have established that *with product fecundities* (in the sense of (1.10)), *random mating is equivalent to random union of gametes*. In particular if only differential viabilities are manifested then the notion of panmictic (= random) mating whether in the guise of random formation of mating types or random union of gametes, is unambiguously defined. However, in the very general situation with no restrictions on the selection parameters, random mating and random union of gametes induce different models.

Some cases of the above result were proposed in various sources (e.g., Kempthorne [1957], Bodmer [1965], Nagylaki and Crow [1974]).

2. REVIEW OF RESULTS ON THE CLASSICAL ONE-LOCUS MULTI-ALLELE VIABILITY SYSTEM

The symmetric (having $M = F$ in (1.3)) multi-allele one-locus viability model follows the transformation equations (cf. (1.4))

$$p_i' = \frac{p_i \sum_{j=1}^{r} m_{ij} p_j}{\sum_{i,j}^{r} m_{ij} p_i p_j}, \quad i = 1, 2, \ldots, r, \tag{2.1}$$

or in our concise notation

$$p' = \frac{p \circ Mp}{(Mp, p)} = Tp \tag{2.2}$$

defined on $p \in \Delta$ (Δ = the simplex of frequency vectors). For this model (referred to henceforth as the *standard one-locus multi-allele model*) the denominator

$$W(p) = \sum_{i,j=1} m_{ij} p_i p_j = (Mp, p) \tag{2.3}$$

called the *fitness function* can be construed as proportional to the mean number of progeny persisting to the next generation, where $(p_1, \ldots, p_r)$ is the allele frequency array of the current generation. It is a remarkable fact that $W(p)$ is a Lyapounov function for the mapping (2.2) (Theorem 2.1 below). This result is most satisfying and consistent with the Darwinian precepts that evolution entails continued fitness improvement. The discovery that $W(p') \geqslant W(p)$ for p and p' related as in (2.1) was apparently first surmised by Fisher and validated rigorously by a number of authors. An elegant proof is found in Kingman [1961a].*

*Several of the properties on equilibria reviewed in this section are due to several authors, notably, Kingman [1961b]; consult e.g., Crow and Kimura [1970], Ewens [1969], for further references; see Karlin [1979] for elaborations and further developments.

THEOREM 2.1 (Wright-Fisher discrete form of the fundamental theorem of natural selection): *For the transformation* (2.2), *the fitness function* $W(p)$ *satisfies*

$$W(p) \leqslant W(Tp) \tag{2.4}$$

with equality prevailing iff $Tp = p$.

The fitness function $W(p)$ serves as a Lyapounov function for the standard one-locus multi-allele model but generally fails in this respect in the context of multi-locus selection models. Because $W(p)$ depicts a quadratic surface, it generally manifests at most a single local maximum interior to Δ. For bisexual one-locus selection models (where in (1.3), $M \neq F$) there exist classes of viability arrays with multiple interior equilibria and therefore any associated Lyapounov function cannot be a quadratic or even a ratio of quadratics. Numerical studies persistently exhibit global convergence of the iterates of the transformation law underlying these models. Nevertheless the determination of a useable Lyapounov function for the general bisexual viability model remains a challenging mathematical problem.

The following elementary characterization of the fixed points of T underlies much of our later developments:

The relation

$$Tp^* = p^*, \quad p^* = (p_1^*, \ldots, p_r^*)$$

holds iff either

$$p_i^* = 0 \text{ or } \sum_j m_{ij} p_j^* = W(p^*), \quad \text{where } p_i^* > 0. \tag{2.5}$$

In another convenient notation, let $I(p^*)$ denote the collection of indices of the positive components of p^*. Then

$$\begin{aligned} \sum_{j \in I(p^*)} m_{ij} p_j^* &= W(p^*), \quad \text{for } i \in I(p^*), \\ p_i^* &= 0, \quad \text{for } i \notin I(p^*). \end{aligned} \tag{2.6}$$

Throughout this section (and also section 3), *unless stated otherwise we impose the following widely applicable nondegeneracy postulate*:

ASSUMPTION 2.1: *Every principal submatrix of the viability matrix* $M = \|m_{ij}\|_{i,j=1}^{r}$ *is non-singular.*

With Assumption 2.1 in force the following complete recipe for calculating the fixed points of T is readily confirmed:

$$p = Tp \text{ iff } \sum_{j=1}^{r} m_{ij}p_j = W(p) \text{ for all } i \text{ where } p_i > 0, \tag{2.7}$$

(cf.(2.5)).

Moreover, for each subset of indices $I \subseteq \{1, 2, \ldots, r\}$ there exists *at most one* fixed point $p(I)$ with the property

$$\begin{aligned} &\sum_{j\in I} m_{ij}\, p_j(I) = W(p) \qquad\qquad \text{for } i \in I \\ &p_i(I) > 0 \text{ if } i \in I, \quad p_i(I) = 0 \text{ if } i \notin I. \end{aligned} \tag{2.8}$$

Thus:

> *Under Assumption* 2.1 *with each index set* $I \subseteq \{1, 2, \ldots, r\}$, *there exists at most one fixed point (equilibrium) computed as in* (2.8). (2.9)

In total there can exist at most $2^r - 1$ equilibria. An equilibrium p^* *involving all alleles* (i.e., where $p_i^* > 0$ for all i) is called a *polymorphism.*

The fact that $W(p)$ is a Lyapounov function for the transformation (2.1) together with the observation (2.9) leads to the basic result:

THEOREM 2.2: *Under Assumption* 2.1, $\lim_{n\to\infty} T^n p$ *exists for every* p.

The following necessary and sufficient condition for local stability of an equilibrium $\hat{p}$ is now easily validated:

THEOREM 2.3: *Let $\hat{p}$ be an equilibrium ($T\hat{p} = \hat{p}$). If $W(\hat{p})$ is a strict local maximum of $W(p)$ (relative to Δ), $\hat{p}$ is a locally stable equilibrium for the transformation T of* (2.2).

The characterization of Theorem 2.3 assures the existence of at least one stable equilibrium located at any point at which $\max_{p \in \Delta} W(p)$ is attained.

The next lemma furnishes a useful sufficient condition that a *boundary* fixed point $\hat{p}$ is *not* locally stable:

LEMMA 2.1: *Let $\hat{p}$ be a boundary fixed point of T involving the index set I. If for some $i_0 \notin I$*

$$\sum_{j \in I} m_{i_0 j} \hat{p}_j > W(p), \tag{2.10}$$

then there exists a neighborhood U of $\hat{p}$ such that for any $p \in \Delta^0$, (Δ^0 denotes the interior of Δ), we have $T^n p \not\to \hat{p}$, that is $T^n p$ does not converge to $\hat{p}$.

For an interior fixed point p^*, of course

$$\sum_j m_{ij} p_j^* = W(p^*) \quad \text{for all } i$$

or written compactly

$$Mp^* = uW(p^*) \text{ (where } u \text{ denotes the vector with all components of value 1).}$$

Pertaining to a polymorphic equilibrium, we have the following information equivalent to the characterization of a local maximum of $W(p)$ at an interior point of Δ:

THEOREM 2.4: *An interior fixed point p^* is locally stable iff*

$$(M\xi, \xi) < 0 \text{ for all } \xi \not\equiv 0, \text{ satisfying } (\xi, u) = 0 \tag{2.11}$$

i.e., *M is negative definite on the linear manifold orthogonal to $u: \mathfrak{M}_u^\perp = \{\xi | (\xi, u) = 0\}$. Moreover, p^* is locally stable iff p^* is globally stable in that $T^n p \to p^*$ for every $p \in \Delta^0$, and*

$$W(p^*) \text{ is a strict maximum of } W(p) \text{ with respect to all } p \in \Delta. \tag{2.12}$$

The next statement embraces known criteria equivalent to (2.11) which are useful in the stability analysis of several concrete examples.

PROPOSITION 2.2: *Let M be a real symmetric matrix with non-negative elements and suppose M is irreducible (meaning that some power M^k has all strictly positive entries; assume also that* det *$(M) \neq 0$ or that M is nonsingular).*

The following are equivalent facts:

(i) *The eigenvalues $\{\lambda_i\}_1^r$ of M satisfy*

$$\lambda_1 > 0 > \lambda_2 \geqslant \lambda_3 \geqslant \cdots \geqslant \lambda_r. \tag{2.13}$$

(ii) *The bordered principal subdeterminants*

$$\Delta_i = \begin{vmatrix} m_{11}, m_{12} & \cdots & m_{1i} \\ m_{21}, m_{2,2} & \cdots & m_{2i} \\ \vdots \quad \vdots & \vdots & \vdots \\ m_{i1}, m_{i,2} & \cdots & m_{i,i} \end{vmatrix} \qquad i = 1, 2, \ldots, r$$

strictly alternate in sign such that

$$(-1)^{i-1}\Delta_i > 0. \tag{2.14}$$

If we assume also the existence of vectors $w \gg 0$, $v \gg 0$ (*all components of* w *and* v *are positive*) *such that* $Mw = v$, *then* (iii) *below is equivalent to* (i).

(iii) $(M\xi, \xi) < 0$ *for all* $\xi \neq 0$ *lying in the hyperplane* $(\xi, v) = 0$.

The equivalence of (ii) and (iii) emanates from the Sylvester Inertia Theorem on symmetric matrices and the Jacobi representation of a quadratic form.

Consider now the *gradient map* $T'(p)$ (i.e., the local linear approximation to T at p). We find for p interior to Δ

$$T'(p)\xi = \frac{\xi \circ Mp}{W(p)} + \frac{p \circ M\xi}{W(p)} - \frac{2(Mp, \xi)}{[W(p)]^2}\, p \circ Mp. \tag{2.15}$$

When $p = p^*$ is an internal fixed point of T (i.e., p^* is a polymorphism), then (2.15) reduces to

$$T'(p^*)\xi = \xi + \frac{p^* \circ M\xi}{W(p^*)} \tag{2.16}$$

for ξ restricted to satisfy $(\xi, u) = 0$. The constraint $(\xi, u) = 0$ is pertinent, as then for ϵ small enough $p^* + \epsilon\xi$ persists as a frequency vector. It is easily verified that the manifold $\mathfrak{M}_u^{\perp}$ is transformed by $T'(p^*)$ into itself. This is a valid general assertion emanating from the fact that p^* is a fixed point of T. The eigenvalues of $T'(p^*)$ in (2.16) are $1 + \gamma_i$, $i = 1, 2, \ldots, r$, where γ_i are the eigenvalues of the matrix $U = p^* \circ M/W(p^*)$. Moreover, $\gamma_1 > |\gamma_i|$, $i = 2, \ldots, r$, holds owing to the Frobenius theorem on positive matrices. Since $u \cdot U = u$, $u = (1, 1, \ldots, 1)$, it follows that $\gamma_1 = 1$ with u its associated left eigenvector. Global stability (with respect to Δ^0) or convergence of $T^n p$ to p^* for all $p \in \Delta^0$ is guaranteed if $|1 + \gamma_i| < 1$ or $\gamma_i < 0$ for $i \geqslant 2$.

We conclude this section with some interesting classes of viability models amenable to a more thorough analysis.

Example I. Let

$$M = aI + bE\ (b > 0, a + b > 0), \tag{2.17}$$

where E is the matrix displaying all entries of value 1 and I is the identity matrix. In this example all homozygotes bear the identical viability value $a + b$ while the heterozygotes carry the common viability value b. A direct verification establishes the internal equilibrium

$$p^* = \left(\frac{1}{r}, \frac{1}{r}, \ldots, \frac{1}{r}\right). \tag{2.18}$$

The eigenvalues of M are $a + br, a, a, \ldots, a$, the value a appearing $r - 1$ times.

Invoking the result of (2.13) we may conclude that p^* *is stable* (globally) *iff* $a < 0$. We find that where (2.18) is unstable then only the pure homozygote states (vertex points of Δ) are locally stable equilibria. In the latter event, if $p = (p_1, \ldots, p_r)$ is the initial frequency configuration and say $p_{i_0} > \max(p_1, \ldots, p_{i_0-1}, p_{i_0+1}, \ldots, p_r)$, then $T^n p$ converges to the pure state $\hat{p}_{i_0} = 1$, $\hat{p}_i = 0$, $i \neq i_0$.

Example II. *Model of multiplicative allelic effects*

Assume that corresponding to allele A_i is an intrinsic viability effect v_i, $i = 1, 2, \ldots, r$ and suppose the viability of the A_iA_j genotype is the independent multiplicative contributions of its allelic constituents such that the viability of the genotype $A_iA_j = m_{ij} = v_iv_j$. In this case the matrix $M = \|m_{ij}\|$ is singular, actually of rank 1. The transformation equations (2.1) reduce to

$$p_i' = \frac{v_i p_i}{(v, p)} \quad i = 1, \ldots, r. \tag{2.19}$$

The outcome is as follows:

Let $v_{i_0} > v_i$ *for all* $i \neq i_0$. *Then from any initial* $p \in \Delta^0$, *fixation of allele* A_{i_0} *ultimately occurs.* (2.20)

Example III. *Model of additive allelic effects*

In this formulation $m_{ij} = v_i + v_j$. The conclusion of (2.20) again applies.

Example IV. *A Lorentzian-Gramian matrix*

The next example was founded on some phenomena of biological compartment theory. Consider r linearly independent real vectors in Euclidean m-space $E^m (m \geqslant r)$:

$$x^\nu = \left(x_1^{(\nu)}, x_2^{(\nu)}, \ldots, x_m^{(\nu)}\right), \nu = 1, 2, \ldots, r, \tag{2.21}$$

and define the Lorentzian inner product of two vectors $y = (y_1, \ldots, y_r)$ and $z = (z_1, \ldots, z_r)$ as

$$((y, z)) = y_1 z_1 - \sum_{i=2}^{m} y_i z_i. \tag{2.22}$$

Consider the matrix

$$M = \|((x^{(\nu)}, x^{(\mu)}))\|_{\nu, \mu=1}^{r} \tag{2.23}$$

which we call a Lorentzian-Gramian matrix.

If a viability matrix M has a representation of the form (2.23), then we can establish the existence of a *unique stable equilibrium*, not necessarily a polymorphism. Stated more precisely, we have:

Let M be defined as in (2.23) *satisfying the conditions of Assumption* 2.1. *Suppose M is generic in the sense that for any equilibrium* $\hat{p}(I)$

$$\sum_{j \in I} m_{ij} \hat{p}_j(I) \neq W(\hat{p}) \text{ for } i \notin I. \tag{2.24}$$

(*Apart from the presence of special relations among the $x^{(\nu)}$ vectors, the fact of* (2.24) *usually prevails.*) *Then there exists a unique stable equilibrium* (*not necessarily a polymorphism*).

Example V. *Some general examples for the existence of a polymorphism* (*internal equilibrium*) *not necessarily stable and connections to game theory*

For some background material on game theory we refer to Karlin [1959]. A matrix game (pay-off matrix M) is said to be "completely mixed" if there exists a frequency vector $x^0 = (x_1^0, \ldots, x_r^0) \gg 0$ and $y^0 = (y_1^0, \ldots, y_r^0) \gg 0$ such that $Mx^0 = u\kappa = y^0 M$ (recall that u is the unit vector) where κ is a constant and M is *non singular*. Obviously where M is symmetric, we have $x^0 = y^0$ and $\kappa = W(x^0)$. A matrix engendering a completely mixed game admits a unique polymorphism $p^* = x^0$. It should be emphasized that the polymorphism is not necessarily stable but we merely affirm its existence.

The information contained in the existence of a polymorphism is of two kinds: (i) Either the polymorphism is locally stable and therefore globally stable, or (ii) there exist *at least two* (usually more) locally stable boundary equilibria.

We display two classes of viability matrices having an interior equilibrium derived from the fact that they induce completely mixed games (consult *loc.cit.* [1959, Chap. 2] for details).

(a) *Minkowski-Leontief matrices are completely mixed.*

A square matrix M is called a Minkowski-Leontief matrix if

(i) $m_{ii} > \gamma \geqslant 0$; (ii) $m_{ij} \leqslant \gamma$ for $i \neq j$; (iii) $\Sigma_{i=1}^{r} m_{ij} > r\gamma$.

(b) *A matrix induced by an ordered set game is completely mixed.*

Let $L_1, L_2, \ldots, L_r$ be disjoint closed sets on the positive real axis obeying the ordering $L_1 > L_2 > \cdots > L_r$. The notation $L_i > L_j$ signifies that all the points of L_i lie to the right of the points of L_j. Let M be the symmetric matrix of the form

$$M = \left\| \begin{array}{cccccc} L_2 & L_3 & L_4 & \cdots & \cdot & L_1 \\ L_3 & L_4 & & \cdots & L_1 & L_2 \\ L_4 & \cdot & \cdot & \cdots & L_2 & L_3 \\ \vdots & \vdots & & \iddots & \iddots & \vdots \\ \cdot & L_1 & L_2 \cdots & & & \vdots \\ L_1 & L_2 & L_3 \cdots & & & L_r \end{array} \right\|,$$

where the element m_{ij} is a value contained in the L-set appearing in the (i, j) place. Algebraically:

$$m_{ij} \in L_{i+j} \text{ if } i + j \leqslant r, \; m_{ij} \in L_{i+j-r} \text{ if } i + j \geqslant r + 1.$$

(c) *Circulants*: We cite finally the example of a circular matrix which always involves the fixed point $p^* = (\frac{1}{r}, \frac{1}{r}, \ldots, \frac{1}{r})$. A symmetric circular matrix has the representation

$$M = \left\| \begin{matrix} m_1 & m_2 & \cdots\cdots & m_{r-1}, & m_0 \\ m_2 & m_3 & & m_0 & m_1 \\ \vdots & & & \vdots & \vdots \\ m_0 & m_1 & \cdots\cdots & m_{r-2} & m_{r-1} \end{matrix} \right\|,$$

where m_i are all positive.

3. FURTHER RESULTS ON THE ONE-LOCUS MULTI-ALLELE SYMMETRIC VIABILITY MODEL

The results reported in this section are all new and largely motivated by questions connected with multi-locus theory.

I. A viability matrix induced by a polymorphism. The construction and theorem presented below are relevant to the multi-locus induced fitness overdominance principle discussed in Section 8 later.

Let p^* be a stable polymorphic equilibrium for the viability matrix M and let p^* be decomposed into k blocks of components corresponding to indices $I_1, I_2, \ldots, I_k$. By relabeling the alleles we can assume without imposing any restrictions that the indices in each group are consecutive

$$\begin{aligned} &I_1 = (1, 2, \ldots, \alpha_1), I_2 = (\alpha_1 + 1, \ldots, \alpha_2), \ldots, \\ &I_k = (\alpha_{k-1} + 1, \ldots, \alpha_k) \qquad (\alpha_k = r) \end{aligned} \tag{3.1}$$

and form the induced vectors of lengths $\alpha_1, \alpha_2, \ldots, \alpha_k$ as follows:

$$q_\nu^* = (p^*_{\alpha_{\nu-1}+1}, p^*_{\alpha_{\nu-1}+2}, \ldots, p^*_{\alpha_\nu}),$$
$$\nu = 1, 2, \ldots, k(\alpha_0 = 0 \text{ by convention}).$$

We next construct the normalized vector (components adding to 1)

$$\tilde{q}_\nu = \left(\frac{p^*_{\alpha_{\nu-1}+1}}{|p^*|_\nu}, \frac{p^*_{\alpha_{\nu-1}+2}}{|p^*|_\nu}, \ldots, \frac{p^*_{\alpha_\nu}}{|p^*|_\nu} \right)$$

where

$$\sum_{i=\alpha_{\nu-1}+1}^{\alpha_\nu} p_i^* = |p^*|_\nu, \tag{3.2}$$

so that $\tilde{q}_\nu$ is a frequency vector of $\alpha_\nu - \alpha_{\nu-1}$ components with indices from I_ν.

Consider the associated decomposition of M in terms of the index blocks yielding

$$M = \begin{Vmatrix} M_{11} & M_{12} & \cdots & M_{1k} \\ \vdots & \vdots & & \vdots \\ M_{k1} & M_{k2} & \cdots & M_{kk} \end{Vmatrix}, \tag{3.3}$$

where

$$M_{\mu\nu} = \begin{Vmatrix} m_{\alpha_{\mu-1}+1,\, \alpha_{\nu-1}+1}, & m_{\alpha_{\mu-1}+1,\, \alpha_{\nu-1}+2} & , \ldots, m_{\alpha_{\mu-1}+1,\, \alpha_\nu} \\ \vdots & \vdots & \vdots \\ m_{\alpha_\mu,\, \alpha_{\nu-1}+1}, & m_{\alpha_\mu,\, \alpha_{\nu-1}+2} & , \ldots, m_{\alpha_\mu,\, \alpha_\nu} \end{Vmatrix}$$

is of order $(\alpha_\mu - \alpha_{\mu-1}) \times (\alpha_\nu - \alpha_{\nu-1})$.

We form the induced $k \times k$ matrix whose elements consist of inner products formed from the matrices $M_{\mu\nu}$ applied to the vectors $\{\tilde{q}_\lambda\}_1^k$

$$V = \|(\tilde{q}_\mu, M_{\mu\nu}\tilde{q}_\nu)\|, \mu, \nu = 1, \ldots, k. \tag{3.4}$$

THEOREM 3.1: *Let p^* be a polymorphic equilibrium for M. Then the vector of k components*

$$x^* = (|p^*|_1, |p^*|_2, \ldots, |p^*|_k) \text{ (consult the notation of (3.2))}$$

is a polymorphism for the matrix V. If, in addition, p^* is stable then x^* is stable.

It is convenient and pertinent to our later deliberations on multi-locus theory to single out the special case $k = 2$.

COROLLARY 3.1: *Let p^* be a stable polymorphism for M and construct the vectors*

$$x = \left[\frac{p_1^*}{\sum_{i=1}^{k} p_i^*}, \frac{p_2^*}{\sum_{i=1}^{k} p_i^*}, \ldots, \frac{p_k^*}{\sum_{i=1}^{k} p_i^*} \right] \text{ and}$$

$$y = \left[\frac{p_{k+1}^*}{\sum_{i=k+1}^{r} p_i^*}, \ldots, \frac{p_r^*}{\sum_{i=k+1}^{r} p_i^*} \right] \tag{3.5}$$

and

$$M = \begin{Vmatrix} M_{11} & M_{12} \\ M_{21} & M_{22} \end{Vmatrix},$$

$$M_{11} = \|m_{ij}\|_1^k,\ M_{22} = \|m_{ij}\|_{k+1}^r,\ M_{12} = \|m_{ij}\|_{i=1, j=k+1}^{k, r} = M_{21}.$$

Then

$$(x, M_{12}y) > \max\left[(x, M_{11}x), (y, M_{22}y)\right].$$

II. Bounds on the number of stable equilibria that can coexist. It is an important fact that the equilibrium picture of any multi-locus model with relatively tight linkage relationships is governed by the fine information accessible from the corresponding multi-allele one-locus theory. For example, the selection consequences of a two-locus two-allele model with zero recombination rate are equivalent to a four-allele one-locus model. Similarly, a two-locus model involving p and q allelic alternatives at the respective loci, for no recombination, reduces to a pq multi-allele model at a single locus. In fact, suppose the alleles at the first locus are $A_1, A_2, \ldots, A_p$ and those at the second locus $B_1, B_2, \ldots, B_q$. The gamete (chromosomal) types relative to both loci constitute the array A_iB_j, $i = 1, 2, \ldots, p$; $j = 1, \ldots, q$. A genotype carrying two chromosomal types is described by

$$\left(\frac{A_iB_j}{A_kB_l}\right),$$

where the combinations A_iB_j are connected on the same chromosome while A_kB_l are linked on the homologous chromosome. Accordingly, the model is equivalent to a one-locus situation involving pq allele alternatives.

By invoking suitably the continuity method of small perturbations (Karlin and McGregor [1972a]), some results pertaining to multi-allele one-locus selection models can be transcribed into a number of properties pertaining to multi-locus models associated with *small* positive recombination rates.

Our key results in assessing bounds on the number of coexisting stable equilibria are as follows:

THEOREM 3.2: *Consider the one-locus r-allele selection model having viability matrix* $M = \|m_{ij}\|$ *with Assumption* 2.1 *in force.*

(i) *There exist at most* $2^r - 1$ *equilibria. More precisely, with each combination of s alleles* ($s \leqslant r$) *there can exist at most one equilibrium involving with positive frequency the prescribed alleles.*

(ii) *If an interior equilibrium exists and is locally stable, then it is globally stable* (*Theorem* 2.1).

(iii) *If there exists a locally stable equilibrium involving exactly* $r - 1$ *alleles* (*so that exactly one of the alleles is not represented*), *then there can exist in total at most* 2 *locally stable equilibria.*

(iv) *When there exists a stable equilibrium involving* $r - k$ *alleles* $(1 \leqslant k \leqslant r-1)$, *then* 2^k *is an upper bound for the total number of coexisting stable equilibria. Moreover, if* $r \geqslant 2k$, *then cases can be constructed where the upper bound* 2^k *is attainable.*

The conclusions of (ii)–(iv) with reference to (i) underscore the fact that most existing equilibria are locally unstable. Unstable equilibria are only of marginal interest in that they partially help to demarcate the domains of attraction for the stable equilibria.

III. Some concepts of heterosis with multiple alleles. The concept of heterozygote advantage (also referred to as "overdominance" and sometimes also as the property of "heterosis") classically applies for one-locus two-alleles signifying that the fitness of A_1A_2 (the heterozygote) exceeds the fitness of both homozygote genotypes A_1A_1 and A_2A_2. Fisher in 1922 observed in the one-locus two-allele context that a globally stable polymorphism exists if and only if heterozygote advantage is in force. It is clearly relevant to ask,"What is an appropriate notion of overdominance (or heterosis) in the context of a one-locus multi-allele system?" As is not uncommon in passing from one to higher dimensions a hierarchy of multi-allelic versions arise that are useful and we indicate three such concepts of increasing strengths.

DEFINITION 3.1: *An* $r \times r$ *viability matrix*

$$M = \|m_{ij}\|_1^r$$

corresponding to r alleles at a single locus is said to be simply heterotic (or simply overdominant) if there exists a stable polymorphism $\hat{x} = (\hat{x}_1, \ldots, \hat{x}_r)$, *all* $\hat{x}_i > 0$.

Equivalently $\hat{x}$ exists satisfying

$$M\hat{x} = (\hat{x}, M\hat{x})u$$

$$(u = \text{the vector displaying all unit components}) \tag{3.6}$$

and the eigenvalues of M obey the relations

$$\lambda_1(M) > 0 > \lambda_2(M) \geqslant \lambda_3(M) \geqslant \cdots \geqslant \lambda_r(M). \quad (3.7)$$

The multi-allele one-locus selection theory (Proposition 2.2 and Theorem 2.2) informs us that local stability of a polymorphic equilibrium $\hat{x}$ entails its global stability.

A stronger notion is that of *total heterosis*:

DEFINITION 3.2: *The viability matrix M is said to be totally heterotic (or totally overdominant) if every principal submatrix of M is heterotic in the sense of Definition* 3.1.

Thus, with total heterosis in force, any subcollection of alleles can maintain a stable equilibrium restricted to these alleles which is however unstable with the introduction of any new allele.

Where M is ordered heterotic we infer that for every specification of indices $i = (i_1, i_2, \ldots, i_k)$

$$1 \leqslant i_1 < i_2 < \cdots < i_k \leqslant r,$$

there exists a probability vector η, $\eta_{i_\nu} > 0$, $\Sigma_{\nu=1}^{k} \eta_{i_\nu} = 1$, satisfying

$$\sum_{\nu=1}^{k} m_{i_\mu, i_\nu} \eta_{i_\nu} = (\eta, M\eta), \quad \mu = 1, \ldots, k \quad (3.8a)$$

and

$$\sum_{\nu=1}^{k} m_{i_0, i_\nu} \eta_{i_\nu} > (\eta, M\eta) \quad \text{for all} \quad i_0 \neq i_\mu. \quad (3.8b)$$

The latter relation affirms that the frequency vector η is unstable with respect to the availability of any allelic type different from those corresponding to the indices $i_1, i_2, \ldots, i_k$.

The following characterizes when a viability matrix can be totally heterotic:

THEOREM 3.3: *A matrix* $M = \|m_{ij}\|_{i,j=1}^{r}$ *is totally heterotic if and only if every principal submatrix*

$$\tilde{M} = \|m_{i_\nu, i_\mu}\|, \ \mu, \nu = 1, 2, \ldots, k$$

$$(1 \leqslant i_1 < i_2 < \cdots < i_k \leqslant r) \tag{3.9}$$

has the property that with each $i_0 \neq i_\mu$ *there exists a positive frequency vector* $\gamma^{(0)}$; $(\gamma_\nu^{(0)} > 0, \nu = 1, \ldots, k, \sum_{\nu=1}^{k} \gamma_\nu^{(0)} = 1)$ *depending on M and the choice of* i_0 *such that*

$$m_{i_0, i_\mu} \geqslant \sum_{\nu=1}^{k} \gamma_\nu^{(0)} m_{i_\nu, i_\mu}, \quad \mu = 1, \ldots, k, \tag{3.10}$$

manifesting strict inequality in at least one component.

There is an even stronger notion than *totally heterotic*:

DEFINITION 3.3: *We say that M is strongly totally heterotic if for every principal submatrix*

$$\tilde{M} = \|m_{i_\nu, i_\mu}\|, \quad \nu, \mu = 1, \ldots, k$$

and for every index i_0 *distinct from all* i_μ, $\mu = 1, 2, \ldots, k$, *there exists at least some row in* $\tilde{M}$ (say i_λ) *with the property*

$$m_{i_0, i_\nu} \geqslant m_{i_\lambda, i_\nu}, \quad \nu = 1, \ldots, k \tag{3.11}$$

with strict inequality in some component.

A one-locus viability matrix of some innate biological interest has the form $V = \|v_{ij}\|$ where all heterozygotes independent of its allelic composition carry a common fitness value 1 ($v_{ij} = 1$ for $i \neq j$) and all homozygotes bear the same fitness value α, $v_{ii} = \alpha$ for all i.

The eigenvalues of V (of order $r \times r$) are

$$\begin{cases} \alpha + r - 1, & \text{having simple multiplicity,} \\ \alpha - 1, & \text{occurring with multiplicity } r - 1. \end{cases}$$

For $0 < \alpha < 1$, the fitness regime V is strong totally heterotic with the central frequency state $\hat{x} = (1/r, 1/r, \ldots, 1/r)$ stable. Applications of these concepts occur in Sections 5 and 6.

IV. Upper and lower bounds on the number of alleles necessary to a stable equilibrium. Taking more careful account of the structure of certain classes of fitness matrices we can extract information on the number of alleles involved in a stable equilibrium. Mathematically, we establish upper and lower bounds on the dimension of the face where the stable equilibria are located. In this direction the following simple consequence of Proposition 2.2 is basic for this purpose.

Consider a non-singular fitness matrix M of order $r \times r$ having eigenvalues

$$\lambda_1 \geqslant \lambda_2 \geqslant \cdots \geqslant \lambda_r, \tag{3.12}$$

where this collection includes l positive and $r - l$ negative terms.

THEOREM 3.4: *Given the data of* (3.12), *any stable equilibrium of the one-locus model induced by M can involve at most*

$$r - l + 1 \text{ allele types.} \tag{3.13}$$

(Since M exhibits only positive entries, the Frobenius theory of positive matrices assures that $l \geqslant 1$.)

In particular, the existence of a stable polymorphic equilibrium requires $l = 1$. Where $l = r$, Theorem 3.3 affirms that only the pure homozygous states can be stable. Where $l \geqslant 2$, the only stable equilibria necessarily are located on the boundary of the frequency simplex Δ.

We next examine a fitness matrix of size $m \times m$ ($m = \Pi_{i=1}^{n} r_i$) of the form

$$M = M_1 \otimes M_2 \otimes \cdots \otimes M_n \tag{3.14}$$

arising as a Kronecker product of the fitness matrices M_k of order $r_k \times r_k$, $k = 1, 2, \ldots, n$. (For information on Kronecker products consult Appendix A at the close of this section.)

Matrices of the structure (3.14) feature prominently in the analysis of non-epistatic (non-interactive) multi-locus systems (consult in this respect Sections 5 and 6).

Let the eigenvalues of M_k be arranged in decreasing order

$$\lambda_1^{(k)} \geqslant \lambda_2^{(k)} \geqslant \cdots > \lambda_{r_k}^{(k)}. \tag{3.15}$$

Where each M_k is overdominant (see (3.7) and Definition 3.1), then

$$\lambda_1^{(k)} > 0 > \lambda_2^{(k)} \geqslant \lambda_3^{(k)} \geqslant \cdots \geqslant \lambda_{r_k}^{(k)}$$

and each such set includes one positive and $r_k - 1$ negative eigenvalues.

It is a familiar fact that the set of eigenvalues for M endowed with the Kronecker product form (3.14) consists of all products

$$\mu_{(i_1,\ldots,i_n)} = \prod_{k=1}^{n} \lambda_{\nu_k}^{(k)} \quad \text{for all choices} \quad 1 \leqslant \nu_k \leqslant r_k. \tag{3.16}$$

Consider the general case where the kth collection (3.15) has l_k positive and $r_k - l_k$ negative eigenvalues. An elementary induction establishes that the set (3.16) contains L^+ and $m - L^+ = L^-$ positive and negative eigenvalues, respectively, where for m *even* (m odd)

$$L^+ = \sum_{(\nu_1,\ldots,\nu_p)} \left[l_{\nu_1} l_{\nu_2} \cdots l_{\nu_p} (r_{\mu_1} - l_{\mu_1}) \cdots (r_{\mu_q} - l_{\mu_q}) \right] \tag{3.17}$$

such that the sum extends over all choices of an *odd* (even) number of indices from $(1,\ldots,n)$ and $(\mu_1,\ldots,\mu_q)$ spans the complementary indices to $(\nu_1,\ldots,\nu_p)$ among $(1,2,\ldots,n)$.

It is useful to specialize (3.17) to the case $n = 2$. We obtain

$$\begin{aligned} L^+ &= (r_1 - l_1)(r_2 - l_2) + l_1 l_2, \\ L^- &= l_1(r_2 - l_2) + l_2(r_1 - l_1). \end{aligned} \tag{3.18}$$

With $r_1 = r_2 = \cdots = r_n = r$, $l_1 = l_2 = \cdots = l_n = l$, the formula (3.17) reduces to

$$L^+ = \sum_{\substack{p=\text{odd} \\ (\text{even})}} \binom{n}{p} l^p (r-l)^{n-p}, \quad n \quad \text{even (odd)} \tag{3.19}$$

and in particular for $r = 2$, $l = 1$, we have

$$L^+ = \sum_{p=\text{odd}} \binom{n}{p} = 2^{n-1}, \quad \text{for all } n, \tag{3.20}$$

while for general r, retaining $l = 1$, then

$$L^+ = \sum_{p=\text{odd}} \binom{n}{p} (r-1)^{n-p} = \frac{r^n - (2-r)^n}{2}, \quad n \text{ even.} \tag{3.21}$$

The result of Theorem 3.4 applied to the viability matrix (3.14) establishes that *any stable equilibrium is represented by at most*

$$\left(\prod_{k=1}^{n} r_k \right) - L^+ + 1 = L^- + 1 \tag{3.22}$$

gamete types.

For $n = 2$, the upper bound reduces to $l_1(r_2 - l_2) + (r_1 - l_1)l_2 + 1$. *Where the marginal fitness matrices M_1 and M_2 are overdominant* (compelling $l_1 = l_2 = 1$), the bound above becomes

$$r_1 + r_2 - 1. \tag{3.23}$$

For $r_1 = r_2 = 2$ (two alleles for M_1 and M_2), the bound in (3.23) is 3 which can be attained.

A lower bound is achieved with the aid of

THEOREM 3.5: *If M is as in* (3.14) *and the matrices $M_{i_1}, M_{i_2}, \ldots, M_{i_p}$ are strong totally heterotic, then every stable equilibrium of M entails at least*

$$\max(r_{i_1}, r_{i_2}, \ldots, r_{i_p}) \tag{3.24}$$

positive components.

Feldman *et al.* [1975] treat a special case of two factors in (3.14) with $r_1 = 3$, $r_2 = 2$ having

$$M_1 = \begin{bmatrix} w & 1 & 1 \\ 1 & w & 1 \\ 1 & 1 & w \end{bmatrix}, \quad M_2 = \begin{pmatrix} w & 1 \\ 1 & w \end{pmatrix}.$$

For $0 < w < 1$, M_1 and M_2 are overdominant— strongly totally heterotic. With respect to this example, the bound $r_1 + r_2 - 1 = 4$ given in (3.23) is achieved under the conditions $w + w^2 > 1$, $w < 1$, while all stable equilibria involve 3 gametes for $w + w^2 < 1$.

APPENDIX A. KRONECKER PRODUCTS OF MATRICES.

We record for ready reference several familiar facts concerning Kronecker products of matrices.

DEFINITION A.1: *Kronecker product of matrices.*

Let A be an $m \times n$ real matrix and B of order $l \times k$. The Kronecker product of A and B, written $A \otimes B$, is constructed explicitly as the partitioned matrix of size $ml \times nk$ of the form

$$A \otimes B = \begin{Vmatrix} a_{11}B & a_{12}B \dots a_{1n}B \\ a_{21}B & a_{22}B \dots a_{2n}B \\ \vdots & \\ a_{ml}B & a_{m2}B \dots a_{mn}B \end{Vmatrix}. \tag{A.1}$$

The operation $A \otimes B$ obeys the distributive and associative laws and commutes with scalar multiplication. More specifically, we have

$$(A + B) \otimes C = A \otimes C + B \otimes C, A \otimes (B + C)$$
$$= A \otimes B + A \otimes C,$$
$$(A \otimes B) \otimes C = A \otimes (B \otimes C), (\gamma A) \otimes B$$
$$= A \otimes (\gamma B) = \gamma(A \otimes B),$$

for any real γ.

The next concept is a specialization of Definition A.1:

DEFINITION A.2: *The Kronecker product of two vectors a and b.*

Let $a = (a_1, a_2, \ldots, a_m)$ and $b = (b_1, b_2, \ldots, b_n)$ be real vectors of m and n components, respectively. The Kronecker product of a and b written $a \otimes b$ is the Kronecker product of the $1 \times n$ and $1 \times m$ matrices induced by the vectors a and b, respectively. In component form, we have

$$a \otimes b = (a_1 b_1, a_1 b_2, \ldots, a_1 b_n, a_2 b_1, \ldots, a_m b_1, a_m b_2, \ldots, a_m b_n). \tag{A.2}$$

The next fact is useful in a variety of applications.
For A and C $m \times m$ real matrices, B and D $n \times n$,

$$(A \otimes B) \cdot (C \otimes D) = AC \otimes BD. \tag{A.3}$$

In particular, let A be an $m \times m$ matrix, B an $n \times n$ matrix, a and a' m-vectors, b and b' n-vectors. Then

$$\begin{aligned} (A \otimes B)(a \otimes b) &= Aa \otimes Bb, \\ (a \otimes b, a' \otimes b') &= (a, a')(b, b'), \end{aligned} \tag{A.4}$$

where (x, y) denotes the scalar product of the two vectors x and y.
Let ϕ be a polynomial in two variables ξ and η with real coefficients

$$\phi(\xi, \eta) = \sum_{i,j=0}^{p} c_{ij} \xi^i \eta^j.$$

The following fact used in our analysis of the multiplicative selection model details the relationship between the eigenvalues of A, B and

$$\phi(A, B) = \sum_{i,j=0}^{p} c_{ij} A^i \otimes B^j. \tag{A.5}$$

THEOREM A.1: *If* $\lambda_1, \lambda_2, \ldots, \lambda_m$ *are the eigenvalues of the* $m \times m$ *matrix* A *and* $\mu_1, \mu_2, \ldots, \mu_n$ *the eigenvalues of the* $n \times n$ *matrix* B, *then the eigenvalues of the matrix* $\phi(A, B)$ *defined in* (A.5) *consist of the* mn *numbers* $\phi(\lambda_r, \mu_s)$, $r = 1, 2, \ldots, m$; $s = 1, 2, \ldots, n$.

COROLLARY A.1:
(i) *The eigenvalues of* $A \otimes B$ *are the* mn *numbers*

$$\lambda_r \mu_s,\ r = 1, 2, \ldots, m;\ s = 1, 2, \ldots, n.$$

(ii) *Moreover, if* $Aa = \lambda a$ *and* $Bb = \mu b$, *then the relation*

$$(A \otimes B)(a \otimes b) = \lambda\mu(a \otimes b) \text{ holds.} \tag{A.6}$$

Using property (ii) we can delineate a complete set of eigenvectors for the Kronecker product matrix $A \otimes B$. Explicitly, suppose the existence of a complete set of eigenvectors $a^{(1)}, \ldots, a^{(m)}$ for the matrix A corresponding to the eigenvalues, $\lambda_1, \ldots, \lambda_m$, respectively, and assume $b^{(1)}, \ldots, b^{(n)}$ exist having a similar meaning with respect to B for the eigenvalues $\mu_1, \mu_2, \ldots, \mu_n$. Then a complete basis of eigenvectors for $A \otimes B$ consist of all Kronecker products

$$a^{(i)} \otimes b^{(j)},\ i = 1, \ldots, m;\ j = 1, \ldots, n. \tag{A.7}$$

DEFINITION A.3: *Schur product of vectors.*

Let a and a' be two m real vectors. The Schur product of a and a', written $a \circ a'$, is defined to be the vector whose components are the pointwise multiplication of the components of a and a', viz.,

$$a \circ a' = (a_1 a'_1, a_2 a'_2, \ldots, a_m a'_m).$$

The following property intertwines the Schur product and the Kronecker product: Let a and a' be m-component real vectors, b

and b' n-component real vectors. Then:

$$(a \otimes b) \circ (a' \otimes b') = (a \circ a') \otimes (b \circ b').$$

The extension of the concept of Schur product to the case of matrices is straightforward.

4. THE ANALYSES OF SOME BISEXUAL MULTI-ALLELE SELECTION SYSTEMS

The formulation of the multi-allele selection model is set forth in Section 1. The transformation equations (1.1) are repeated here for ready reference

$$p' = \frac{\frac{1}{2}[p \circ Fq + q \circ Fp]}{(p, Fq)}, \quad q' = \frac{\frac{1}{2}[p \circ Mq + q \circ Mp]}{(p, Mq)}, \tag{4.1}$$

or more compactly $\{p', q'\} = T\{p, q\}$,

where p and q are frequency vectors of the allele types in the female and male populations, respectively. Unlike the symmetric case ($F = M$) we know of no Lyapounov function or way to determine all the equilibria of the system (4.1) apart from the case with $r = 2$ (involving two alternative alleles), although computer runs and numerical work imply global convergence for the iterates T^n. As three polymorphic equilibria (at most two stable) can occur for $r = 2$, it follows that no quadratic function or ratio of two quadratics can generally serve as a Lyapounov function in the bisexual model. There appears to be no simple analog of the mean fitness function that featured fundamentally in the discussion of the one-sex viability model (or symmetric bisexual case). In the analyses of the general bisexual model, ad hoc devices and ingenuity are our only recourse.

The theory for the bisexual model is mathematically challenging and there are numerous unresolved problems. In this section we shall develop the stability analysis of equilibria for some important subclasses. Partial results for the case $r = 2$ are found in Owens [1953], Bodmer [1965], among others. In the general case, consult Cannings [1969b]. The examples treated in this section are mostly new.

Let $\{\hat{p}, \hat{q}\}$ be a polymorphic fixed point of (4.1), i.e., $\hat{p}_i\hat{q}_j > 0$ for all i, j. The gradient matrix at $\{\hat{p}, \hat{q}\}$ can be determined standardly to obtain for $T'\{\hat{p}, \hat{q}\}$ the matrix mapping

$$\begin{aligned} \xi' &= C_{F,\hat{q}}\xi - \frac{\hat{p}\,(\xi, F\hat{q})}{(\hat{p}, F\hat{q})} + C_{F,\hat{p}}\eta - \frac{\hat{p}\,(\eta, F\hat{p})}{(\hat{p}, F\hat{q})}, \\ \eta' &= C_{M,\hat{q}}\xi - \frac{\hat{q}\,(\xi, M\hat{q})}{(\hat{p}, M\hat{q})} + C_{M,\hat{p}}\eta - \frac{\hat{q}\,(\eta, M\hat{p})}{(\hat{p}, M\hat{q})}, \end{aligned} \tag{4.2}$$

where

$$C_{F,\hat{q}}\xi = \frac{1}{2}\,\frac{\xi \circ F\hat{q} + \hat{q} \circ F\xi}{(\hat{p}, F\hat{q})}, \tag{4.3}$$

with the matrix transformation $C_{F,\hat{p}}$, $C_{M,\hat{q}}$, $C_{M,\hat{p}}$ determined analogously. We can write the linear approximation near $\{\hat{p}, \hat{q}\}$ succinctly as

$$\begin{pmatrix} \xi' \\ \eta' \end{pmatrix} = \begin{pmatrix} U_1 & U_2 \\ V_1 & V_2 \end{pmatrix}\begin{pmatrix} \xi \\ \eta \end{pmatrix}, \tag{4.4}$$

where U_i and V_i are appropriate $r \times r$ matrices that depend on $\{\hat{p}, \hat{q}\}$ and the viability matrices F and M (e.g., $U_1\xi = C_{F,\hat{q}}\xi - (\xi, F\hat{q})\hat{p}/(\hat{p}, F\hat{q})$). The equilibrium $\{\hat{p}, \hat{q}\}$ is locally stable if the magnitude of the eigenvalues, for the transformation (4.4) on the invariant subspace $\mathfrak{M} = \{\xi, \eta : (\xi, u) = (\eta, u) = 0\}$ are less than unity. It is useful to record some facts concerning (4.4) in the special case $U_1 = U_2$ and $V_1 = V_2$ such that the gradient matrix transformation has the form

$$\begin{pmatrix} \xi' \\ \eta' \end{pmatrix} = \begin{pmatrix} U & U \\ V & V \end{pmatrix}\begin{pmatrix} \xi \\ \eta \end{pmatrix} = \Gamma\begin{pmatrix} \xi \\ \eta \end{pmatrix}. \tag{4.5}$$

LEMMA 4.1: *Consider a matrix mapping of the form* (4.5), *with* ξ *and* η *in* E^r. *The eigenvalues of* Γ *are those of* $S = U + V$ *plus the eigenvalue* 0 *which occurs with additional multiplicity* r.

An important class of cases engendering a local linear approximation of the kind (4.5) arises as follows. Suppose the polymorphic equilibrium $\{\hat{p}, \hat{q}\}$ satisfies $\hat{p} = \hat{q}$ and

$$F\hat{p} = W_F(\hat{p})u, \quad M\hat{q} = W_M(\hat{q})u, \tag{4.6}$$

where $W_F(p) = (p, Fp)$ is the fitness function determined by F, and accordingly $W_M(q) = (q, Mq)$. If $(\xi, u) = 0$ and $(\eta, u) = 0$, then (4.6) yields, because $\hat{p} = \hat{q}$,

$$(\xi, F\hat{q}) = 0, \quad (\eta, F\hat{p}) = 0, \quad (\xi, M\hat{q}) = 0, \quad (\eta, M\hat{p}) = 0.$$

Also the equality $C_{F,\hat{p}} = C_{F,\hat{q}}$ obviously holds, and

$$C_{F,\hat{q}}\xi = \frac{\frac{1}{2}\xi \circ F\hat{q} + \hat{q} \circ F\xi}{(\hat{p}, F\hat{q})} = \frac{1}{2}\xi + \frac{1}{2}\,\frac{\hat{p} \circ F\xi}{W_F(\hat{p})} = U\xi. \tag{4.7}$$

In the same way, we find $C_{M,\hat{p}} = C_{M,\hat{q}}$ with

$$C_{M,\hat{q}}\xi = \frac{1}{2}\xi + \frac{1}{2}\,\frac{q \circ M\xi}{W_M(\hat{q})} = V\xi.$$

We now treat some specific examples.

Example I. *Viability matrices satisfying the relation* $M = aE + bF$, $a, b \geqslant 0$ *with general* F *where* $E = \|e_{ij}\|_1^r$, $e_{ij} = 1$ *for all* i, j.

If there exists p^* such that $Fp^* = \gamma u$, p^* a positive frequency vector with $\gamma > 0$, then $\{p^*, p^*\}$ is a polymorphic equilibrium for the bisexual transformation associated with F and $M = aE + bF$

and the gradient matrix $T'\{p^*, p^*\}$ has the form (4.5) with

$$U\xi = \frac{1}{2}\xi + \frac{1}{2}\,\frac{p^* \circ F\xi}{W_F(p^*)} \quad \text{and} \quad V\xi = \frac{1}{2}\xi + \frac{1}{2}\,\frac{p^* \circ M\xi}{W_M(p^*)}\,.$$

Further $W_M(p^*) = a + bW_F(p^*)$.

In order to check the stability of $\{p^*, p^*\}$ we need to determine the eigenvalues of

$$(U + V)\zeta = I\zeta + \frac{1}{2}\,\frac{p^* \circ F\zeta}{W_F(p^*)} + \frac{1}{2}\,\frac{p^* \circ (aE + bF)\zeta}{a + bW_F(p^*)} \tag{4.8}$$

on the manifold of all ζ satisfying $(\zeta, u) = 0$.

A direct analysis of (4.8) reveals that a sufficient condition for the polymorphism $\{p^*, p^*\}$ to be locally stable is that it satisfy the inequality

$$\left| 1 + \left(\frac{1}{2} + \frac{bW_F(p^*)}{2(a + bW_F(p^*)} \right) \lambda \right| < 1. \tag{4.9}$$

for all eigenvalues λ different from unity of the matrix transformation $X\xi = \dfrac{p^* \circ F}{W_F(p^*)}\,\xi$. If (4.9) is reversed for one of the λ, then $\{p^*, p^*\}$ is not stable.

Two particular cases of interest are $a = 0$ and $b = 0$.

Case (i). $a = 0$ or $M = bF$, so that selection operates equally in the two sexes and the formulation can be viewed as a one-sex multi-allele model of the kind already dealt with in Section 2. The condition for stability of the polymorphic equilibrium $\{p^*, p^*\}$ becomes

$$|1 + \lambda| < 1 \text{ satisfied for all eigenvalues } \lambda \text{ distinct from 1 of the matrix}, \ \Gamma\xi = \frac{p^* \circ F\xi}{W_F(p^*)}\,. \tag{4.10}$$

CASE (ii). $b = 0$ or $M = aE$, where selection for the trait is manifested only in the female while the male shows no differential selection effects between the genotypes. The condition for stability of $\{p^*, p^*\}$ now becomes

$$\left|1 + \frac{\lambda}{2}\right| < 1. \tag{4.11}$$

A plot of the condition (4.11) contrasted to (4.10) in the complex plane is helpful:

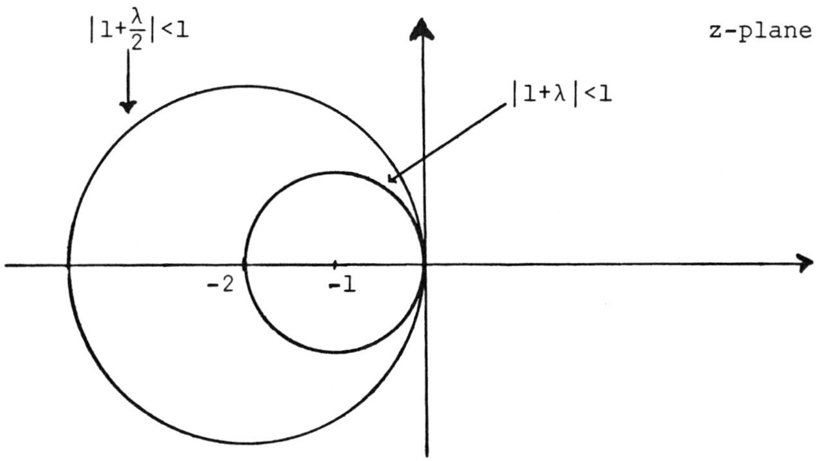

FIG. 1.

Comparing the two cases, with differential selection acting only on female genotypes vis-a-vis where both sexes manifest the same selection regime, we see that in the second case there is greater likelihood for $\{p^*, p^*\}$ to be stable.

Example II. $F = \|f_{ij}\|$, $M = \|m_{ij}\|$, *with* $f_{ij} = b$ *and* $m_{ij} = a$ *for* $i \neq j$ *and* $f_{ii} = m_{ii} = 1$.

Here all homozygotes carry a common viability value and heterozygotes have an identical fitness value but these values differ between the sexes. We can represent both F and M in the form

$$F = bE + (1 - b)I,$$
$$M = aE + (1 - a)I, \tag{4.12}$$

(recall $E = \|e_{ij}\|$ with $e_{ij} \equiv 1$).

This example is a particular case of Example I, but now a complete elaboration of its fixed points and their stability properties are available. The polymorphism $p_i^* = q_i^* = \frac{1}{r}$, $i = 1, 2, \ldots, r$, always exists while simple analysis shows that other polymorphic equilibria if extant must satisfy the relations

$$\hat{q}_i = \frac{\left(B - \frac{b}{2}\right)\hat{p}_i}{(1 - b)\hat{p}_i + b/2}, \quad i = 1, 2, \ldots, r, \tag{4.13}$$

where

$$B - \frac{b}{2} = \frac{b}{2} + (1 - b)\sum_{i=1}^{r} \hat{p}_i\hat{q}_i = \frac{1}{2}\,\frac{a}{a - 1}(1 - b).$$

Paraphrasing the discussion of Example I, we can ascertain that a sufficient condition for $\{p^*, q^*\}$ to be locally stable is the fulfillment of the inequality

$$\left| \frac{\dfrac{b}{2} + \dfrac{1 - b}{r}}{b + \dfrac{1 - b}{r}} + \frac{\dfrac{a}{2} + \dfrac{1 - a}{r}}{a + \dfrac{1 - a}{r}} \right| < 1. \tag{4.14}$$

Three cases need to be distinguished:

(i) $a, b < 1$: Concomitantly, it is easily checked that (4.14) is impossible and $\{p^*, q^*\}$ is locally unstable. This outcome is intuitive.

(ii) $a, b > 1$: We confirm the natural result that $\{p^*, q^*\}$ is locally stable.

(iii) $a > 1 > b$: A rearrangement of (4.14) reveals its validity iff

$$\frac{1}{r} < \frac{1}{2}\left[\frac{a}{a-1} - \frac{b}{1-b}\right], \tag{4.15}$$

and then $\{p^*, q^*\}$ is locally stable (in fact globally stable).

We discuss next the case when the inequality in (4.15) is reversed. It is then appropriate to examine the relevance of a polymorphism of the kind obeying the relations (4.13). Notice that when $a > 1 > b$, then equilibria of the form (4.13) are feasible iff there exists a frequency solution $\{\hat{p}_i\}$ of

$$S(\hat{p}) = \sum_{i=1}^{r} \frac{\hat{p}_i}{(1-b)\hat{p}_i + \frac{b}{2}} = \frac{2(a-1)}{a(1-b)}.$$

It is elementary to check that $S(p)$ describes a concave surface whose maximum

$$\frac{1}{\frac{b}{2} + \frac{1-b}{r}}$$

is achieved at

$$p^* = \left(\frac{1}{r}, \ldots, \frac{1}{r}\right).$$

It follows that the domain defined by $S(p) = 2(a-1)/a(1-b)$ intersects the interior of the simplex $\Delta = \{p = (p_1, \ldots, p_r) : p_i \geqslant 0, \Sigma p_i = 1\}$ iff $a > 1 > b$ and r satisfies

$$\frac{1}{\frac{b}{2} + \frac{1-b}{r}} \geqslant \frac{1}{\frac{1}{2}\frac{a}{a-1}(1-b)} > \frac{1}{1-\frac{b}{2}}. \tag{4.16}$$

The final inequality occurs iff $(a+b)/2 > 1$. A straightforward analysis reveals that the corner equilibria $\tilde{P}_{A_1} = (1, 0, \ldots, 0)$, $\tilde{P}_{A_2} = (0, 1, 0, \ldots, 0)$, etc., are locally stable iff $(a+b)/2 < 1$.

The surface of polymorphisms can be calculated from $S(\hat{p}) = 2(a-1)/a(1-b)$ and (4.13). Where they exist, they are locally stable. The following pictures (in the case $r = 3$, $a > 1 > b$) depict the various possibilities. The arrows indicate the convergence directions of $T^n p$ for the case at hand.

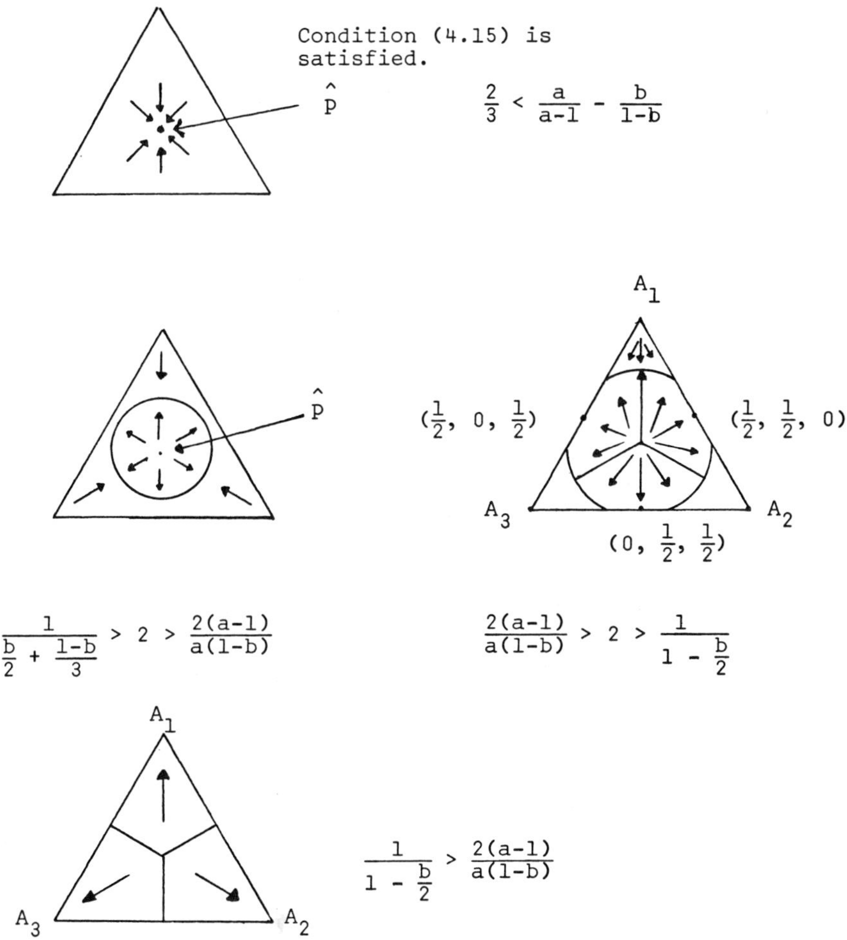

FIG. 2.

Example III.

$$F \text{ is a diagonal matrix } \begin{cases} f_{ij} = 0, & i \neq j, \\ f_{ii} = f_i > 0, & i = 1, 2, \ldots, r, \end{cases}$$

$$M \text{ is a diagonal matrix } \begin{cases} m_{ij} = 0, & i \neq j, \\ m_{ii} = m_i > 0, & i = 1, 2, \ldots, r. \end{cases}$$

In this example all heterozygote genotypes are lethal or near lethal.

A single polymorphism $\{\hat{p}, \hat{q}\}$ exists whose components are

$$\hat{p}_i = \frac{1}{m_i} \Big/ \sum_{j=1}^{r} \frac{1}{m_j}; \quad \hat{q}_i = \frac{1}{f_i} \Big/ \sum_{j=1}^{r} \frac{1}{f_j}.$$

This polymorphism is not locally stable and in fact convergence to an appropriate corner point which depends on the initial frequency vector always occurs.

More specifically, for an initial p, q if we determine i_0 as the index satisfying

$$m_{i_0} f_{i_0} p_{i_0} q_{i_0} = \max_{1 \leqslant i \leqslant r} (m_i f_i p_i q_i),$$

where we assume for simplicity that i_0 is unique, then for all $i \neq i_0$, $p_i^{(n)} \underset{n\to\infty}{\longrightarrow} 0$ and $q_i^{(n)} \underset{n\to\infty}{\longrightarrow} 0$ and of course $p_{i_0}^{(n)} \underset{n\to\infty}{\longrightarrow} 1$ and $q_{i_0}^{(n)} \underset{n\to\infty}{\longrightarrow} 1$.

Example IV. *The additive case* $f_{ij} = f_i + f_j$; $m_{ij} = m_i + m_j$.

We stipulate that the coefficients f_i and m_i are in "general position" meaning that for each triple $\{i, j, k\}$,

$$\begin{vmatrix} 1 & f_i & m_i \\ 1 & f_j & m_j \\ 1 & f_k & m_k \end{vmatrix} \neq 0.$$

Under this condition it can be shown that for any equilibrium there can exist at most two components with $\hat{p}_i \hat{q}_i > 0$. The case of two is attainable already where two alternate alleles are possible.

Example V. *The multiplicative case* $f_{ij} = f_i f_j$; $m_{ij} = m_i m_j$.

If we assume that for each i, j, k, l,

$$\begin{vmatrix} 1 & f_i & m_i & m_i f_i \\ 1 & f_j & m_j & m_j f_j \\ 1 & f_k & m_k & m_k f_k \\ 1 & f_l & m_l & m_l f_l \end{vmatrix} \neq 0,$$

then each equilibrium can involve at most three alleles and generally only two alleles.

5. SOME SELECTION MODELS INVOLVING TWO-LOCUS TRAITS

Consider a character determined by two loci with possible alleles A, a and B, b at the first and second locus, respectively. There are therefore four gamete types, also referred to as chromosomal types:

$$AB \quad Ab \quad aB \quad ab \tag{5.1}$$

and 10 genotypes (individual zygote types)

$$\frac{AB}{AB}, \frac{AB}{Ab}, \frac{Ab}{Ab}, \frac{AB}{aB}, \frac{AB}{ab}, \frac{Ab}{aB}, \frac{Ab}{ab}, \frac{aB}{aB}, \frac{aB}{ab}, \frac{ab}{ab},$$

where the symbol AB/aB, for example, means that the alleles A and B sit on a common chromosome while the alleles a and B are found on the homologous chromosome. Let $M = \|w_{ij}\|_{i,j=1}^{4}$ denote the fitness matrix, where w_{ij} is the fitness of the genotype composed from the i and j chromosome types.

A phenomenon of considerable importance to the maintenance of genetic variability is the mechanism of recombination (at meiosis) that is widely identified in the case of multi-locus traits. Thus with recombination a pair of loci are not independent so far as gamete donation is concerned. For example, an individual of genotype AB/ab can produce gametes of type AB and ab and also chromosomal types Ab and aB. When all four are produced in equal numbers the loci are called unlinked or free. The AB and ab gametes are called parental types while the Ab and aB gametes are called recombinant types. If the loci are linked there will be an excess of parental over recombinant gametes. It is generally found that the parental types AB, ab are produced with equal frequencies $\frac{1}{2}(1 - r)$ and the recombinant types with equal frequencies $\frac{1}{2}r$ where the number r, $0 < r \leqslant 1$, (usually $r \leqslant \frac{1}{2}$) is called the *recombination rate* or *frequency*. For the physical explanation of the phenomenon and more details on its importance the reader may consult any genetics text.

Let x_1, x_2, x_3, x_4 be the frequencies of the four gamete types in the order of (5.1). Assuming random union of gametes (random mating) and Mendelian segregation involving recombination frequency r, we find that the frequency x_1' of AB in the next generation is proportional to

$$\begin{aligned} x_1' &\sim x_1 w_{11} + 2x_1 x_2 w_{12} \tfrac{1}{2} + 2x_1 x_3 w_{13} \tfrac{1}{2} \\ &\quad + 2x_1 x_4 w_{14}(1 - r)\tfrac{1}{2} + 2x_2 x_3 r \tfrac{1}{2} w_{23} \\ &= x_1 w_1 - rD, \end{aligned}$$

where $w_1 = \sum_{j=1}^{4} w_{1j} x_j$, $D = x_1 x_4 w_{14} - x_2 x_3 w_{23}$. Similar expressions result for x_2', x_3' and x_4'. Henceforth, for ease of exposition we assume $w_{14} = w_{23}$.

The transformation equations relating the gamete frequencies over two successive generations can be compactly expressed in the form

$$x' = \frac{x \circ Mx - erD(x, x) w_{23}}{W(x)} = Tx, \tag{5.2}$$

where $x = (x_1, x_2, x_3, x_4)$, $e = (1, -1, -1, 1)$, and

$$W(x) = \sum_{i,j} w_{ij} x_i x_j = (x, Mx); \tag{5.3}$$

$D(x) = D(x, x)$ is the bilinear form $(x, Dx) = x_1 x_4 - x_2 x_3$ induced by the matrix

$$D=\frac{1}{2}\begin{Vmatrix}0&0&0&1\\0&0&-1&0\\0&-1&0&0\\1&0&0&0\end{Vmatrix}=\frac{1}{2}\tilde{D}\otimes\tilde{D},\ \tilde{D}=\begin{Vmatrix}0&1\\-1&0\end{Vmatrix}. \qquad (5.4)$$

It is convenient for later purposes to display the genotype fitness matrix:

$$\begin{array}{c} \\ AB\\ Ab\\ aB\\ ab\end{array}\begin{array}{c}\begin{array}{cccc}AB & Ab & aB & ab\end{array}\\ \begin{Vmatrix}w_{11}&w_{12}&w_{13}&w_{14}\\w_{21}&w_{22}&w_{23}&w_{24}\\w_{31}&w_{32}&w_{33}&w_{34}\\w_{41}&w_{42}&w_{43}&w_{44}\end{Vmatrix}\end{array}, \qquad (5.5)$$

where, for example, w_{34} is the fitness of the zygote aB/ab composed of chromosomes 3 and 4.

Another familiar display of the genotype viability regime has the form

$$\begin{array}{cccc} & & 2nd\ locus & \\ 1st\ locus & BB & Bb & bb \\ AA & w_{11} & w_{12} & w_{22} \\ Aa & w_{13} & \{w_{23}, w_{14}\} & w_{24} \\ aa & w_{33} & w_{34} & w_{44} \\ & & viabilities & \end{array} \qquad (5.6)$$

The ambiguity $\{w_{23}, w_{14}\}$ arises since A/a at locus one and B/b at locus two can be combined in the coupling form AB/ab or repulsion form Ab/aB. Henceforth, we assume the coupling and repulsion heterozygote fitnesses are equal to 1; the normalization 1 is proper since only relative fitness values are pertinent.

According to this assumption the expression $D = D(x) = x_1x_4 - x_2x_3$ is usually dubbed the *linkage disequilibrium function* evaluated at the frequency state x. When $D = 0$, the population is then said to be in linkage equilibrium.

If we write $p_1 = x_1 + x_2$ as the frequency of the "A" gene and $p_2 = x_1 + x_3$ as that of the "B" gene, then

$$x_1 = p_1p_2 + D, \quad x_2 = p_1(1 - p_2) - D,$$
$$x_3 = p_2(1 - p_1) - D, \quad x_4 = (1 - p_1)(1 - p_2) + D,$$

and clearly at linkage equilibrium the frequency of a chromosome is just the product of the frequencies of its constituent genes. D might therefore be regarded as one possible measure referring to the departure of a chromosome from being in a state of random association of its genes.

Three classes of viability patterns have been most prominently studied. (Contributors in this regard include Wright, Kimura, Bodmer, Felsenstein, Ewens, Feldman, Lewontin, Franklin, Allard, Jain, McGregor, Karlin and many others; see Karlin [1975] for references and a review of two-locus theory.) These are:

(i) additive fitnesses
(ii) multiplicative fitnesses
} representing non-epistasis (meaning no interaction of fitness effects between the separate loci)

and

(iii) the symmetric viability model.

If the fitness array (5.6) admits the representation:

Fitnesses at locus 1 \ Fitnesses at locus 2	BB β_1	Bb β_2	bb β_3
AA α_1	$w_{11} = \alpha_1 + \beta_1$	$w_{12} = \alpha_1 + \beta_2$	$w_{22} = \alpha_1 + \beta_3$
Aa α_2	$w_{13} = \alpha_2 + \beta_1$	$w_{14} = \alpha_2 + \beta_2 = w_{23}$	$w_{24} = \alpha_2 + \beta_3$
aa α_3	$w_{33} = \alpha_3 + \beta_1$	$w_{34} = \alpha_3 + \beta_2$	$w_{44} = \alpha_3 + \beta_3$,

(5.7)

then we say there is no additive epistasis.

Where (5.6) takes the form:

Fitnesses at locus 1 \ Fitnesses at locus 2	BB β_1	Bb β_2	bb β_3
AA α_1	$w_{11} = \alpha_1\beta_1$	$w_{12} = \alpha_1\beta_2$	$w_{22} = \alpha_1\beta_3$
Aa α_2	$w_{13} = \alpha_2\beta_1$	$w_{14} = \alpha_2\beta_2$	$w_{24} = \alpha_2\beta_3$
aa α_3	$w_{33} = \alpha_3\beta_1$	$w_{34} = \alpha_3\beta_2$	$w_{44} = \alpha_3\beta_3$,

(5.8)

then there would be no multiplicative epistasis. The existence of epistasis in essence means that the fitness of a genotype cannot be partitioned into independent contributions due to each locus separately, in either of the two senses of (5.7) or (5.8).

The final class of fitness schemes for which some explicit results have been derived to date is the symmetric viability model where the fitnesses are related in the form:

$$\begin{array}{cc} & \begin{array}{ccc} BB & Bb & bb \end{array} \\ \begin{array}{c} AA \\ Aa \\ aa \end{array} & \begin{bmatrix} w_{11} = \alpha & w_{12} = \beta & w_{22} = \delta \\ w_{13} = \gamma & w_{14} = \varepsilon = w_{23} & w_{24} = \gamma \\ w_{33} = \delta & w_{34} = \beta & w_{44} = \alpha \end{bmatrix} \end{array} \tag{5.9}$$

This fitness array exhibits the property that the viability of a genotype is unaltered by the simultaneous relabelings $A \to a$, $a \to A$, $B \to b$, and $b \to B$ so that, e.g., AB/aB and ab/Ab carry the same fitness value.

For the models (5.7) and (5.8) it is meaningful to refer to the *marginal fitnesses* associated with each separate locus while for (5.9) the notion of a marginal fitness is not well defined. Consonantly, the underlying selection effects of the individual loci cannot be disentangled when dealing with a general viability scheme except for the non-epistatic systems. This does not preclude the outcome that the observed chromosomal frequency array behaves as independent contributions from separate loci gene frequency factors at equilibrium. We discuss this issue more in section 6.

It is instructive to display the transformation equations for models (5.7)–(5.9) in terms of Kronecker products of matrices. The symmetric forms in which the recombination parameter appears attest to the intrinsic nature of these formulas.

Representation for the two-locus multiplicative model. Set

$$M_1 = \begin{Vmatrix} \alpha_1 & \alpha_2 \\ \alpha_2 & \alpha_3 \end{Vmatrix}, \qquad M_2 = \begin{Vmatrix} \beta_1 & \beta_2 \\ \beta_2 & \beta_3 \end{Vmatrix}, \tag{5.10}$$

corresponding to the one-locus viability matrices implicit in (5.8). The transformation equation (5.2) for the case (5.8) possesses the canonical representation as a sum of simple Kronecker products (for the notation, see Appendix A at the close of section 3):

$$\begin{aligned} W(x)x' = {} & (1-r)(I \otimes I)x \circ (M_1 \otimes M_2)x \\ & + r(M_1 \otimes I)x \circ (I \otimes M_2)x, \end{aligned} \tag{5.11}$$

where $W(x) = (x, M_1 \otimes M_2 x)$ and I is the identity matrix.

Representation for the two-locus additive model. If M denotes the additive fitness matrix corresponding to (5.7), then we have the representation

$$M = M_1 \otimes E + E \otimes M_2, \tag{5.12}$$

where $E = \begin{Vmatrix} 1 & 1 \\ 1 & 1 \end{Vmatrix}$ and M_1 and M_2, the marginal fitness matrices, are given as in (5.10). The expression of (5.2) associated with (5.12) is the natural extension of (5.11) for a sum. Specifically, the transformation equation assumes the form

$$\begin{aligned} W(x)x' = {} & (1-r)(I \otimes I)x \circ (M_1 \otimes E)x \\ & + r(M_1 \otimes I)x \circ (I \otimes E)x \\ & + (1-r)(I \otimes I)x \circ (E \otimes M_2)x \\ & + r(E \otimes I)x \circ (I \otimes M_2)x, \end{aligned} \tag{5.13}$$

where $W(x) = (x, Mx) = (x, M_1 \otimes Ex) + (x, E \otimes M_2 x)$.

Representation for the two-locus symmetric model. Let us introduce the matrices

$$J = \begin{Vmatrix} 0, & 1 \\ 1, & 0 \end{Vmatrix} \quad \text{and} \quad K = \begin{Vmatrix} 1, & 0 \\ 0, & -1 \end{Vmatrix}. \tag{5.14}$$

The fitness matrix M corresponding to (5.9) possesses the representation

$$M = \frac{\alpha + \delta}{2} I \otimes I + \left(\frac{\alpha - \delta}{2} \right) K \otimes K + \beta J \otimes I + \gamma I \otimes J + \varepsilon J \otimes J. \tag{5.15}$$

The analog of (5.13) for (5.15) is achieved by adding the transformations of the type (5.11) applied to each term of (5.15).

The advantage of the representation (5.15) resides in the fact that the vectors $u = (1, 1)$ and $e = (1, -1)$ constitute eigenvectors for the matrices J, E and I. Moreover, we obviously have $Ku = e$ and $Ke = u$. For this model it is very convenient to operate in terms of the basis vectors $u \otimes u$, $u \otimes e$, $e \otimes u$, $e \otimes e$. We will amply illustrate this approach in Section 6.

Results for the two-locus multiplicative and additive models have been set forth by a number of authors. When the contributions of the loci to the viabilities are additive, there is a single interior polymorphism which is globally stable for all non-zero recombination values when both loci are heterotic (Karlin and Feldman [1970a]). The equilibrium population is in linkage equilibrium which means that the population gamete frequencies factor as the products of the separate loci gene frequencies. When the viabilities are multiplicative over loci, it is known that loose linkage coupled to heterozygote advantage at the separate loci is sufficient for global stability of the equilibrium displaying linkage equilibrium (Moran [1968], Bodmer and Felsenstein [1967]). For this model with sufficiently small recombination values, Karlin and McGregor [1972] have established the existence of two stable polymorphic equilibria in strong linkage disequilibrium. Concerning further recent results on these special classical selection models and also for general viability regimes, see Karlin [1975].

In the remainder of this section we highlight new results concerning the theory of two-locus selection balance whose full development and further ramifications are set forth in Karlin [1979].

Invariant regions in the multiplicative model: Two loci, two alleles. Consider the frequency domain

$$\Delta = \left\{ x = (x_1, x_2, x_3, x_4) | x_i \geqslant 0, \quad \sum_{i=1}^{4} x_i = 1 \right\}$$

and define

$$R_+ = \{x | D(x) = x_1 x_4 - x_2 x_3 > 0\}, \quad R_- = \{x | D(x) < 0\}. \tag{5.16}$$

The separating surface $D(x) = 0$ consists of all frequency states in *linkage equilibrium*. The following theorem holds:

THEOREM 5.1: (i) *For $r \leqslant \frac{1}{2}$ the regions R_+ and R_- are invariant under the transformation* (5.11), *i.e., the sign of $D(x)$ is preserved provided $r \leqslant 1/2$.* (ii) *Moreover, if $D(x) = 0$, then $D(x') = 0$ where this property holds for all r.*

Recall that $D(x) = \frac{1}{2}(x, \tilde{D} \otimes \tilde{D}x)$ where $\tilde{D} = \left\| \begin{matrix} 0, & 1 \\ -1, & 0 \end{matrix} \right\|$. It is not difficult to establish the inequality

$$D(x') = (\beta(x) - r) D(x) \quad \text{with} \quad \beta(x) \geqslant \frac{1}{2}$$

from which the assertions of Theorem 5.1 ensue.

Some facets of the two-locus multiplicative viability model with multiple alleles. The representation (5.11) extends immediately to the case where locus 1 involves p alleles having marginal viability matrix $M_1 = \|m_{ij}^{(1)}\|_{i,j=1}^{p}$ and the second locus has q alleles with marginal viability matrix $M_2 = \|m_{ij}^{(2)}\|_{i,j=1}^{q}$. If we label the alleles at locus 1 by $A_1, A_2, \ldots, A_p$ and those at locus 2 by $B_1, B_2, \ldots, B_q$, then the two-locus system involves pq chromosomal types A_iB_j which we arrange for convenience in lexicographic order. Let $x = \{x_{ij}\}$ be a frequency vector of the chromosomal types. The transformation equation in this pq frequency space reads identically to (5.11), viz.,

$$x' = \frac{\begin{array}{c}(1-r)(I \otimes I)x \circ (M_1 \otimes M_2)x \\ +r(M_1 \otimes I)x \circ (I \otimes M_2)x\end{array}}{(x, M_1 \otimes M_2 x)} = Tx. \qquad (5.17)$$

It is elementary to verify that if $x^{(1)}$ is a polymorphic equilibrium for M_1 and $x^{(2)}$ for M_2, then the Kronecker product vector

$$x^* = x^{(1)} \otimes x^{(2)} \qquad (5.18)$$

is an equilibrium for T referred to as a Hardy-Weinberg (H. W.) point owing to the property that the x^* component frequency of A_iB_j under (5.18) is calculated as the product of its marginal gene frequencies. It can be shown that the equilibrium (5.18) can never be locally stable for absolute linkage (when $r = 0$) and therefore for small recombination. It, therefore, is of some interest to describe some results on the nature of the stable equilibria of (5.17) when $r = 0$. The developments on orders of heterosis for multi-allele one-locus models set forth in paragraph IV of Section 3 are relevant here. We highlight one concomitant consequence.

Consider a viability matrix of the form

$$M = M_1 \otimes M_2, \qquad (5.19)$$

where M_1 and M_2 as characterized previously correspond to the one-locus fitness matrices of p and q alleles at locus 1 and 2, respectively. *Suppose M_1 is strongly heterotic* (see Definition 3.3).

Let $\hat{x} = (\hat{x}_{ij})$ be a stable equilibrium for (5.19). We assert that every allele A_k at locus 1 is involved in $\hat{x}$ in that there is some component of x where

$$\hat{x}_{k,j} > 0 \quad \text{for some} \quad j.$$

Moreover, we have

THEOREM 5.2: *Consider a fitness matrix of the form* (5.19) *with* $p(q)$ *possible alleles at locus* 1 (2). *Suppose* M_1 *and* M_2 *are both strongly heterotic. Then every stable equilibrium comprises an* ***allelic polymorphism**** (*meaning that every possible allele at locus* 1 *and at locus* 2 *is involved with positive frequency*).

On the other hand, not every chromosomal type occurs with positive frequency since the upper bound of the number of chromosomal types in any stable equilibrium is $p + q - 1$ (*compare to* (3.23)). (*The available number of chromosomal types is* pq.)

Since every allele is necessarily represented, the lower bound of the number of chromosomal types involved in a stable equilibrium is

$$\max(p,q) \tag{5.20}$$

in accordance with Theorem 3.5.

As an application we consider the two-locus system with alleles $A_1, A_2, \ldots, A_p$ possible at locus 1 and $B_1, B_2, \ldots, B_q$ possible at locus 2. Assume the viability of all pure homozygotes is c, i.e., $m\left(\dfrac{A_iB_k}{A_iB_k}\right) = c$ independent of i and k; all single heterozygotes at locus 1 have $m\left(\dfrac{A_iB_k}{A_jB_k}\right) = b,\ i \neq j$; while at locus 2, $m\left(\dfrac{A_iB_k}{A_iB_l}\right) = a,\ k \neq l$, and all double heterozygotes have common fitness value 1.

Assuming the alternative chromosomal types are arranged in lexicographic order, viz., $A_1B_1, A_1B_2, \ldots, A_1B_q, A_2B_1, \ldots,$ then a natural representation of the viability matrix fulfilling the prescriptions stated, is

$$M = V_a \otimes V_b + (c - ab)I \otimes I, \tag{5.21}$$

*This is now common terminology. A chromosomal polymorphism, more than an allelic polymorphism, signifies that every chromosomal type occurs with positive frequency.

where V (of the proper size) is defined to be

$$V_a = \begin{Vmatrix} a & 1 & 1 & \cdots & 1 \\ 1 & a & 1 & \cdots & 1 \\ 1 & 1 & a & \cdots & 1 \\ \vdots & \vdots & \vdots & \ddots & \vdots \\ 1 & 1 & 1 & \cdots & a \end{Vmatrix}.$$

With $a < 1$, the fitness matrix V_a is strong totally heterotic in the sense of Definition 3.3.

Under the specialization $c = ab, 0 < a, \ b < 1$, the fitness matrix (5.21) reduces to the form (5.19) with both factors strongly heterotic. Application of Theorem 5.2 implies that any stable equilibrium displays an allelic polymorphism and entails at least $\max(p, q)$ but at most $p + q - 1$ chromosomal types. In particular, for $p = 2$, q general, any stable equilibrium involves at least q and at most $q + 1$ chromosomal types. These bounds are sharp.

Returning to the general case, the eigenvalues of (5.21) are readily determined to be

$$\begin{array}{ll} [a + (p - 1)][b + (q - 1)] + c - ab & \text{simple multiplicity} \\ [a + (p - 1)](b - 1) + c - ab & \text{multiplicity } q - 1 \\ (a - 1)(b + q - 1) + c - ab & \text{multiplicity } p - 1 \\ (a - 1)(b - 1) + c - ab & \text{multiplicity } (p - 1)(q - 1). \end{array} \tag{5.22}$$

An upper bound on the number of chromosomal types involved in any stable equilibrium is the cumulative number of negative values in (5.22) (counting multiplicity) plus one. For the model with fitness matrix (5.21) the central equilibrium with frequencies

$$\hat{x}_{ij} = \frac{1}{pq} \quad \text{for all } i, j \tag{5.23}$$

is globally stable for all recombination rates provided a, b and c

satisfy

$$(a + p - 1)(b - 1) + c - ab < 0$$

$$(b + q - 1)(a - 1) + c - ab < 0$$

and

$$(a - 1)(b - 1) + c - ab < 0.$$

With $a, b, c < 1$, *then every stable equilibrium is necessarily an allelic polymorphism although the chromosomal polymorphism* (5.22) *is not stable.*

A result on the two-locus multiple-allele additive fitness model. Let $A_1, A_2, \ldots, A_p$ be the possible alleles at the first locus and $B_1, B_2, \ldots, B_q$ those at the second locus. When the effects of the alleles at each of the two loci, on the relative fitnesses of the genotypes, are additive, then the fitness parameter of a typical genotype A_iB_k/A_jB_l is expressed as the sum of fitness parameters of A_iA_j, and B_kB_l. Thus, where $U = \|u_{ij}\|_1^p$ and $V = \|v_{kl}\|_1^q$ are the (marginal) fitness matrices associated with the first and second locus, respectively, then under additive selection contributions the fitness matrix M of the two-locus system can be represented as

$$M = U \otimes E + E \otimes V, \tag{5.24}$$

where E is a matrix of exclusive unity elements.

(We should write $M = U \otimes E_1 + E_2 \otimes V$, where E_1 is $p \times p$ and E_2 is $q \times q$, but we omit the indices as no confusion is conceivable.)

The transformation $x' = Tx$ connecting chromosomal frequencies over two successive generations with recombination rate r has the form (cf. (5.13))

$$W(x)x' = (1 - r)x \circ (U \otimes E + E \otimes V)x$$
$$+ r(U \otimes I)x \circ (I \otimes E)x + r(E \otimes I)x \circ (I \otimes V)x, \tag{5.25}$$

where $W(x) = (x, U \otimes Ex + E \otimes Vx) = (x, Mx)$ *is the mean fitness function.*

Suppose p^* is a stable polymorphism for U and q^* a stable polymorphism for V. The following theorem holds (developed in collaboration with Lieberman):

THEOREM 5.3: *Under the conditions stated on p^* and q^* immediately above and any recombination rate $r > 0$, the frequency vector $x^* = p^* \otimes q^*$ is the unique globally stable polymorphism of the transformation* (5.25).

This extends a corresponding result of Karlin and Feldman [1970a] entailing two alleles at each locus. The result of Theorem 5.3 undoubtedly extends to the multi-locus (more than 2) additive fitness model, but no proof is available as yet.

General results on two-locus polymorphisms. We conclude this section by summarizing a few results on the nature of the polymorphic equilibria of (5.2) involving a general viability regime not restricted to be of the special forms encompassed by (5.7)-(5.9). Consider a polymorphic equilibrium $\hat{x}(r)$ (that is $T\hat{x} = \hat{x}$ holds) displaying its dependence on the recombination parameter r.

We next underscore some relevant ancillary facts pertinent to the stability nature of these equilibrium points.

(i) *An equilibrium $\hat{x}(r)$ is independent of r if and only if its linkage disequilibrium value, $D(\hat{x})$, vanishes.* (ii) *With respect to the basic frequency simplex Δ, the gradient matrix $T'(\hat{x}) = A$, operating on the vector ζ, has the form*

$$A\zeta = \frac{\zeta \circ M\hat{x} + \hat{x} \circ M\zeta - 2reD(\hat{x}, \zeta) - 2r\hat{x}D(\hat{x})\left(\frac{e}{\hat{x}}, \zeta\right)}{W(\hat{x})}, \tag{5.26}$$

where we confine the domain $\mathfrak{M}_u^{\perp}$ of A to consist of all ζ satisfying $(\zeta, u) = 0$, $u = (1, 1, 1, 1)$ (we employed the notation $e = (1, -1, -1, 1)$ and $\frac{e}{x} = \left(\frac{e_1}{x_1}, \frac{e_2}{x_2}, \frac{e_3}{x_3}, \frac{e_4}{x_4}\right)$); $D(\hat{x}, \zeta)$ is the bilinear form induced by the matrix of (5.4); it is elementary to check that $\mathfrak{M}_u^{\perp}$ is invariant under A.

It can be shown that if $\hat{x}(r)$ is geometrically locally stable on the segment $r_1 < r < r_2$, then $d\hat{x}(r)/dr$ (abbreviate to $\dot{x}(r)$) exists continuously. With this in hand we can prove the following theorem describing the variation of the mean fitness function $dW(\hat{x}(r))/dr$ as the recombination parameter increases over $r_1 < r < r_2$:

THEOREM 5.4: *Let the viability matrix M be non-singular. Suppose there exists a non-degenerate curve $\hat{x}(r)$ of stable polymorphisms for r ranging in a segment $0 < r < r_0$ all expressing linkage disequilibrium $D(\hat{x}(r), \hat{x}(r)) \neq 0$. Provided r_0 is small enough, there exists r_1 such that*

$$\dot{W}(r) = \frac{d}{dr} W(\hat{x}(r)) < 0, \quad 0 < r < r_1, \quad (r_1 \leqslant r_0). \tag{5.27}$$

Moreover, if $\hat{x}(0)$ is a stable polymorphism, then

$$\dot{W}(0) = 0. \tag{5.28}$$

The result of Theorem 5.4 can be sharpened for the symmetric viability regime of (5.9).

THEOREM 5.5: *Let $\hat{x}(r)$, $r_1 < r < r_2$, be a non-degenerate curve of symmetric stable polymorphisms for the viability model* (5.9). *Then*

$$\dot{W}(r) = \frac{d}{dr} W(\hat{x}(r)) < 0, \quad r_1 < r < r_2. \tag{5.29}$$

(Notice in (5.29) that the interval location (r_1, r_2) is not restricted in any way.)

In general, $\dot{W}(r) < 0$ is not valid in an arbitrary two-locus viability regime along a curve of stable equilibria $\hat{x}(r)$ where r traverses moderate to free recombination values (see Karlin and Carmelli [1975b]).

The findings of Theorems 5.4 and 5.5 play an important role in the study of the evolution of secondary loci that modify effects at other loci. For example, in recent years, considerable effort has been put forth in the investigation of the fate of modifier alleles which modify linkage between other loci, e.g., see Nei [1967], Feldman [1972], Karlin and McGregor [1974] and others. These authors mainly deal with an infinite randomly mating population in which there is a pair of loci subject to selection with reproduction by random mating. A secondary locus (or actual third locus in the specific genetic system at hand) controls the recombination fraction between the two primary loci while generally not directly altering or affecting the selection regime operating on the controlled loci. The study of this kind of secondary selection relies on the facts of Theorems 5.4 and 5.5.

6. RESULTS AND FACETS OF MULTI-LOCUS NONEPISTATIC SELECTION MODELS

Early work in quantitative genetics developed around a single gene. Those single-locus results are often applied in the multi-locus case by adding or multiplying over loci. The two-locus selection theory establishes that the *Hardy-Weinberg* (H. W.) *property* (= linkage equilibrium) does not hold in general. Even when a H. W. equilibrium exists it does not mean that selection is not operating as is commonly inferred. Present knowledge further indicates that the assumption of two alleles per locus is not always realistic and in fact situations of multiple alleles are abundantly and increasingly documented as biochemical separation techniques are refined. The existence of a Hardy-Weinberg-type stable equilibrium is mostly tied to nonepistatic selection systems for loose linkage or symmetric viability patterns for special parameter specifications.

Roux [1974] provided a significant advance for the study of the stability nature of a H. W. equilibrium in the *multiplicative* multilocus selection regime by recognizing the innate Kronecker product structure of this model. We have substantially extended and refined the methodology to deal with a whole hierarchy of selection patterns.

Feldman *et al.* [1975] studied some aspects of the equilibrium structure of a special two-locus multiplicative, symmetric selection regime with two alleles at one locus and three alleles at the other. They were interested in the problem of determining the effect of asymmetry on the number of alleles at two loci but where selection values were symmetric in that all double heterozygotes have a common fitness value, the single heterozygotes all carry an identical fitness value, and all pure homozygotes are equally fit. Christiansen and Feldman [1975] pursued the investigation to cover some cases with two alleles at one locus and $k(\geqslant 3)$ alleles at the second locus.

Franklin and Lewontin [1970] (for a 36-locus situation) examined, through simulation studies, a special selection regime that is simultaneously *symmetric* with respect to allelic substitutions and *multiplicative* in the effects of the separate loci. They concluded that tight linkage produces stable genotypic configurations with high pairwise linkage disequilibrium, while loose linkage points to a stable polymorphism entailing linkage equilibrium of all orders. The extremely high complementarity is, of course, not robust and crucially depends on the multiplicative nonepistatic postulate for the fitness expression coupled to the symmetry imposed with respect to loci and allele effects. Those authors further suggest a measure of association among genes based on pairwise disequilibrium values. A serious drawback in the emphasis on the linkage disequilibrium function D is that it is principally a two-locus two-allele concept. For the circumstance of multiple alleles at two loci, the proponents of this concept advocate the examination of all pairwise D functions where in each evaluation two alleles (one at each locus) are distinguished and the other alleles are lumped creating a single alternative allele at each locus. The unresolved difficulties in statistical theory attendant to the interpretation of partial correlation coefficients are also encountered here in dealing with this vector of D values. Its weaknesses are magnified when treating multi-locus multi-allele systems. We will discuss some alternative measures of association in a future publication.

The general multi-locus multi-allele selection model: Formulation. We concentrate on a diploid population characterized by a trait determined at n loci with n_k possible alleles at locus k.

Let $i_0^{(k)}, i_1^{(k)}$ designate two of the possible alleles at the kth locus. Thus gametes or chromosomal types are given by n-tuples as now displayed.

$$i_0 = (i_0^{(1)}, i_0^{(2)}, \ldots, i_0^{(n)}) \quad \text{and} \quad i_1 = (i_1^{(1)}, i_1^{(2)}, \ldots, i_1^{(n)}). \tag{6.1}$$

(To ease notational complications we will frequently use the short vector symbol i_0, i_1.) A typical genotype composed of the two gametes of (6.2) below is conveniently described by the symbol

$$\left(\frac{i_0^{(1)}, i_0^{(2)}, \ldots, i_0^{(n)}}{i_1^{(1)}, i_1^{(2)}, \ldots, i_1^{(n)}} \right) = \left(\frac{i_0}{i_1} \right), \tag{6.2}$$

signifying that the allelic makeup at locus k involves $i_0^{(k)}, i_1^{(k)}$.

The *recombination frequencies* are summarized by the array of parameters

$$R(\varepsilon_1, \varepsilon_2, \ldots, \varepsilon_n), \tag{6.3}$$

defined for all choices of $\varepsilon_i = 0$ or 1, $i = 1, 2, \ldots, n$, and these numbers are to be interpreted as follows.

The gametic output from segregation of the genotype (6.2) is $(i_{\varepsilon_1}^{(1)}, i_{\varepsilon_2}^{(2)}, \ldots, i_{\varepsilon_n}^{(n)})$ with frequency $R(\varepsilon_1, \varepsilon_2, \ldots, \varepsilon_n)$. The recombination frequencies are stipulated to obey the following intrinsic relations $R(\varepsilon) = R(\varepsilon_1, \varepsilon_2, \ldots, \varepsilon_n) = R(1 - \varepsilon_1, 1 - \varepsilon_2, \ldots, 1 - \varepsilon_n) = R(1 - \varepsilon)$, expressing the fact that the two parental gametes contribute in a symmetrical manner in the segregation process; $\Sigma_\varepsilon R(\varepsilon) = 1$, where the sum extends over all $\varepsilon = (\varepsilon_1, \varepsilon_2, \ldots, \varepsilon_n)$, $\varepsilon_i = 0$ or 1; $i = 1, 2, \ldots, n$. (The $R(\varepsilon)$ by their definition correspond to mutually exclusive recombination events.)

Example 1. (Two loci, $n = 2$.) If r is the frequency of recombination between the two loci, then

$$R(0, 0) = R(1, 1) = \frac{1 - r}{2}, \; R(0, 1) = R(1, 0) = \frac{r}{2}. \tag{6.4}$$

Example 2. (Three loci, $n = 3$.) Let r be the frequency of the event of recombination between loci 1 and 2 coupled with no recombination between loci 2 and 3. The recombination frequency between loci 2 and 3 entailing no recombination between loci 1 and 2 is measured by the parameter s. Finally, let t be the frequency of simultaneous recombination between loci 1 and 2 and loci 2 and 3. With these definitions we obtain

$$R(0,0,0) = R(1,1,1) = \frac{1-r-s-t}{2}, R(0,0,1)$$
$$= R(1,1,0) = \frac{s}{2},$$
$$R(0,1,1) = R(1,0,0) = \frac{r}{2}, R(0,1,0) = R(1,0,1) = \frac{t}{2}. \tag{6.5}$$

A different parametrization of the recombination rates is more commonly used. Consider the case of 3 loci labeled A, B and C. Suppose r_1 is the recombination fraction between the $\{A\}$ locus and the $\{B\}$ locus, that between the $\{B\}$ locus and the $\{C\}$ locus r_2, and that between the $\{A\}$ and $\{C\}$ loci r_3. The parameters in the two notations are connected by the relations

$$r_1 = r + t, r_2 = s + t, r_3 = r + s. \tag{6.6}$$

The case of *no interference* is characterized by the equation $t = (r + t)(s + t)$ or equivalently by the relation

$$r_3 = r_1(1 - r_2) + r_2(1 - r_1) = r_1 + r_2 - 2r_1r_2. \tag{6.7}$$

The situation where recombination between two adjacent positions precludes recombination in the next segment corresponds to the stipulation $t = 0$.

Fitness parameters. Let

$$m\left(\frac{i_0^{(1)}, \ldots, i_0^{(n)}}{i_1^{(1)}, \ldots, i_1^{(n)}}\right) = m\left(\frac{i_0}{i_1}\right) \tag{6.8}$$

denote the relative fitness value of the genotype (6.2). The array

(6.8), where $i_0 = (i_0^{(1)}, \ldots, i_0^{(n)})$ and $i_1 = (i_1^{(1)}, \ldots, i_1^{(n)})$ traverse all chromosomal possibilities, determines a matrix M of order $\Pi_{k=1}^{n} n_k \times \Pi_{k=1}^{n} n_k$.

The specialization of (6.8) to the multiplicative and additive selection models is now worth noting. For *multiplicative non-epistatic selection* effects we stipulate

$$m\left(\frac{i_0^{(1)}, i_0^{(2)}, \ldots, i_1^{(n)}}{i_1^{(1)}, i_1^{(2)}, \ldots, i_1^{(n)}}\right) = \prod_{k=1}^{n} m^{(k)}\big(i_0^{(k)}, i_1^{(k)}\big), \tag{6.9}$$

where

$$\|m^{(k)}(i,j)\| = \|m_{ij}^{(k)}\|_{i,j=1}^{n_k} \tag{6.10}$$

can be construed as the marginal fitness matrix associated with locus k such that the fitness of the genotype (6.2) is determined by independent (multiplicative) fitness contributions from the separate loci.

For *additive non-epistatic selection* the evaluation on the right of (6.9) replaces product by sum, viz.,

$$m\left(\frac{i_0^{(1)}, \ldots, i_0^{(n)}}{i_1^{(1)}, \ldots, i_1^{(n)}}\right) = \sum_{k=1}^{n} m^{(k)}\big(i_0^{(k)}, i_1^{(k)}\big). \tag{6.11}$$

In matrix terminology, the expression (6.9) is recognized as a Kronecker product. In symbols

$$M = M_1 \otimes M_2 \otimes \cdots \otimes M_n, \tag{6.12}$$

where M_k is the matrix of (6.10).

The representation of (6.12) as a Kronecker product allows an expeditious discussion of several aspects of the general multiplicative selection model.

The transformation equations for the multiplicative model (6.12). Let $x(i_0^{(1)}, i_0^{(2)}, \ldots, i_0^{(n)}) = x(i_0)$ denote the frequency of the chromosomal type $i_0 = (i_0^{(1)}, \ldots, i_0^{(n)})$ in the population in the current generation. Under random union of gametes (equivalent to random mating) with the effects of selection operating, the relative proportion of the genotypes (6.2) among mature individuals is

$$m\left(\frac{i_0^{(1)}, i_0^{(2)}, \ldots, i_0^{(n)}}{i_1^{(1)}, i_1^{(2)}, \ldots, i_1^{(n)}}\right) \cdot x\left(i_0^{(1)}, \ldots, i_0^{(n)}\right) \cdot x\left(i_1^{(1)}, \ldots, i_1^{(n)}\right)$$

$$= m\left(\frac{i_0}{i_1}\right) x(i_0) x(i_1).$$

Finally, the segregation process coupled with recombination leads to the frequency of the gamete type $x'(i_0^{(1)}, \ldots, i_0^{(n)})$ in the next generation as computed by the formula

$$W(x)x'(i_0) = \sum_{i_1} \sum_{\varepsilon} R(\varepsilon) m\left(\frac{i_{\varepsilon_1}^{(1)}, \ldots, i_{\varepsilon_n}^{(n)}}{i_{1-\varepsilon_1}^{(1)}, \ldots, i_{1-\varepsilon_n}^{(n)}}\right)$$

$$\cdot\, x\left(i_{\varepsilon_1}^{(1)}, i_{\varepsilon_2}^{(2)}, \ldots, i_{\varepsilon_n}^{(n)}\right) \cdot x\left(i_{1-\varepsilon_1}^{(1)}, \ldots, i_{1-\varepsilon_n}^{(n)}\right), \tag{6.13}$$

where $W(x)$ is the mean fitness function for the population composition x, namely, the sum of the right side in (6.13) extended over all gamete types $i_0 = (i_0^{(1)}, \ldots, i_0^{(n)})$:

$$W(x) = \sum_{i_0, i_1} m\left(\frac{i_0^{(1)}, \ldots, i_0^{(n)}}{i_1^{(1)}, \ldots, i_1^{(n)}}\right) x(i_0) x(i_1).$$

In the important case where the underlying fitness matrix has multiplicative structure (6.9), then (6.13) can be compactly expressed in terms of Kronecker and Schur products in the form

$$W(x)x' = \sum_{\varepsilon} R(\varepsilon)(M_1^{\varepsilon_1} \otimes M_2^{\varepsilon_2} \otimes \cdots \otimes M_n^{\varepsilon_n})x \circ$$

$$(M_1^{1-\varepsilon_1} \otimes M_2^{1-\varepsilon_2} \otimes \cdots \otimes M_n^{1-\varepsilon_n})x, \tag{6.14}$$

where

$$W(x) = (x, M_1 \otimes M_2 \otimes \cdots \otimes M_n x). \tag{6.15}$$

It is instructive to write the version of (6.14) in the case of three loci (cf. for the two-locus case to (5.17)).

Transformation equation for the three loci multiplicative model.

$$\begin{aligned} W(x)x' = {} & (1 - r - s - t)(I \otimes I \otimes I)x \circ (M_1 \otimes M_2 \otimes M_3)x \\ & + r(M_1 \otimes I \otimes I)x \circ (I \otimes M_2 \otimes M_3)x \\ & + s(M_1 \otimes M_2 \otimes I)x \circ (I \otimes I \otimes M_3)x \\ & + t(M_1 \otimes I \otimes M_3)x \circ (I \otimes M_2 \otimes I)x, \end{aligned} \tag{6.16}$$

where $W(x) = (x, M_1 \otimes M_2 \otimes M_3 x)$.

The recombination parameters r, s, t are defined in (6.5). We stress again that the recombination frequencies r, s, t enter into the transformation equations in a manifest symmetrical setting evident from (6.16). This would not be the case if the traditional parameters r_1, r_2 and r_3 were used.

DEFINITION 6.1: *A vector x of $\Pi_{k=1}^n n_k$ coordinates is said to be of Hardy-Weinberg type (abbreviated* H. W.) *if x admits the Kronecker product representation*

$$x = x_{(1)} \otimes x_{(2)} \otimes \cdots \otimes x_{(n)}, \tag{6.17}$$

where $x_{(k)}$ is a vector of size n_k.

An important fact is now highlighted.

THEOREM 6.1: *Suppose that for each locus the marginal fitness matrix M_k admits a polymorphism z_k (when M_k is non-singular, $\hat{z}_k$ is* *unique), so that*

$$M_k \hat{z}_k = W_k(\hat{z}_k) u_k \quad (u_k \textit{ is the unit vector with } n_k \textit{ components}). \tag{6.18a}$$

Then

$$\hat{z} = \prod_{k=1}^{n} \otimes z_k = z_1 \otimes z_2 \otimes \cdots \otimes z_n \tag{6.18b}$$

is a polymorphism of the n-locus system with multiplicative viability structure (6.12).

There is a converse to the above theorem which we will not state here.

Conditions for stability of Hardy-Weinberg polymorphisms. Let $\hat{x} = \hat{x}_1 \otimes \cdots \otimes \hat{x}_n$ be a H. W. equilibrium. It is correct that the component vector $\hat{x}_k$ is then necessarily a polymorphism for the one-locus multi-allele selection model with fitness matrix M_k. The one locus theory tells us that $\hat{x}_k$ is stable if and only if the eigenvalues $\{\mu_i^{(k)}\}_1^{n_k}$ of the matrix

$$\tilde{M}_k = \frac{\hat{x}_k \circ M_k}{W_k(\hat{x}_k)} \tag{6.19}$$

($\tilde{M}_k$ is the matrix M_k multiplied on the left by the diagonal matrix having the components of the vector $\hat{x}_k / W_k(\hat{x}_k)$ on the diagonal; equivalently $\tilde{M}_k$ acting on the vector ξ produces $x_k \circ M_k \xi / W_k(\hat{x}_k)$) satisfy

$$1 = \mu_1^{(k)} > 0 > \mu_i^{(k)}, \; i = 2, \ldots, n_k. \tag{6.20}$$

Since M_k is a symmetric matrix, it is an elementary fact that the eigenvalues of $\tilde{M}_k$ are real. The fact of $\mu_1^{(k)} = 1$ is inherent to the construction of (6.19). The condition for stability of a H. W. polymorphic equilibrium is the content of the next theorem.

THEOREM 6.2: *The polymorphism* $\hat{x} = \hat{x}_1 \otimes \hat{x}_2 \otimes \cdots \otimes \hat{x}_n$ *is locally stable provided*

$$\left| 2 \sum_{\varepsilon} R(\varepsilon) \lambda_1^{\varepsilon_1} \cdot, \lambda_2^{\varepsilon_2} \cdots \lambda_n^{\varepsilon_n} \right| < 1 \tag{6.21}$$

holds for all choices of $\{\lambda_i\}$, *where* λ_k *is an eigenvalue of* $\tilde{M}_k$, $k = 1, 2, \ldots, n$ *excluding the specification* $\lambda_1 = \lambda_2 = \cdots = \lambda_n = 1$ ($\tilde{M}_k$ *defined in* (6.19)).

This important result was first noted by Roux [1974]. Our analysis can be used to treat a variety of selection models possessing Hardy-Weinberg equilibria including a bisexual extension of Theorems 6.1 and 6.2, the additive fitness analog (see later in this section), mixed additive-multiplicative selection schemes, and other classes of genotypic-phenotypic selection associations.

Multiplicative selection with tight linkage. It can readily be checked that the conditions (6.21) are never fulfilled for absolute linkage (no recombination) where $R(\mathit{1}) = R(1, 1, 1 \ldots 1) = R(0, \ldots, 0) = R(\mathit{0}) = \frac{1}{2}$. This fact also emanates from Theorem 3.4. It is of prime interest to determine the nature of the stable equilibria possibilities for absolute linkage which, of course, prescribes the outcomes for tight linkage.

We consider here the case of multiplicative viability contributions from the separate loci where the marginal fitness matrix at locus k, with p alleles, has the special form

$$M_k = \|m_{ij}^{(k)}\|_{i,j=1}^{p}, \quad \text{where } m_{ij}^{(k)} = \begin{cases} 1, & i \neq j, \\ \alpha_k, & i = j, \\ & k = 1, 2, \ldots, n, \end{cases} \tag{6.22}$$

signifying that the viability has a common value for all heterozygotes and another common value for all homozygotes. (More general forms of the multiplicative model under tight linkage will be delineated elsewhere.)

If $0 < \alpha_k < 1$, then M_k is strongly heterotic (see Definition 3.3).

We cited earlier the simulation computations of Franklin and Lewontin [1970] of some multiplicative symmetric multi-locus models having alleles A_i and a_i possible at locus i with all loci tightly linked.

Let the marginal fitness matrices be given by

$$M_i = \begin{Vmatrix} \alpha_i & 1 \\ 1 & \alpha_i \end{Vmatrix}, \quad 0 < \alpha_i < 1,$$

displaying overdominance at each individual locus. The total system viability matrix is given by

$$M = \prod_{i=1}^{n} \otimes M_i. \tag{6.23}$$

Actually Franklin and Lewontin restricted attention to the case $\alpha_1 = \alpha_2 = \cdots = \alpha_n$ with n specified as 2, 5, 18 and 36; these choices partly dictated by computer capabilities and partly to facilitate the elaborate simulation runs required. These authors extrapolated on the basis of their numerical work that in the presence of tight linkage, a population state evolves manifesting principally two complementary chromosomal types (see below for the precise definition), while all other chromosomal types occur with a minute frequency maintained by the recombination process.

We can describe the array of chromosomal types associated with the fitness scheme M of (6.23) in terms of sequences of 0's and 1's

$$C = (\delta_1, \delta_2, \delta_3, \delta_4, \ldots, \delta_n), \ \delta_i = 0 \text{ or } 1, \tag{6.24}$$

where the coordinate value 1 (0) for the chromosome C signifies the constituent allele A (a) for that locus. *A pair of chromosomal types C_1 and C_2 are said to be complementary* if at every coordinate the two types carry distinct alleles. To illustrate, with $n = 2$ there are two pairs of complementary chromosomal sets

$$\{(1, 1), (0, 0)\} \text{ and } \{(1, 0), (0, 1)\}.$$

With n loci there obviously exist 2^{n-1} pairings of complementary chromosomes.

We have established the following fact:

THEOREM 6.3: (i) *Consider a pair of complementary chromosomal types* C_1 *and* C_2. *The frequency vector* $x(C_1, C_2)$ *which assigns probability* $1/2$ *to each of* C_1 *and* C_2 *and* 0 *to all other chromosomal types is a stable equilibrium for the n-locus fitness matrix* (6.23) *in the presence of absolute linkage.*

There are 2^{n-1} *such stable equilibria states corresponding to all pairs of complementary chromosomes.*

(ii) *The only stable equilibria are those concentrating on two complementary chromosomal types as described in* (i).

The simulations of Franklin and Lewontin suggest that for tight linkage usually an equilibrium involving two complementary chromosomes is established for a random mating selection process with *multiplicative symmetric fitness matrix* entailing any number of loci. This conclusion is affirmed as an analytic fact in Theorem 6.3(ii). A number of the principles expressed by Franklin and Lewontin for the expected equilibrium gamete configurations depend strongly on the type of selection system assumed and their extrapolation to general selection recombination structures is mostly not appropriate.

Theorem 6.3 is established in the context of zero recombination rates. However, the small perturbation theory of Karlin and McGregor [1972a] implies that with small recombination frequencies all stable equilibrium states are near the equilibria positions existing for no recombination. Accordingly, two main complementary chromosomes occur with frequencies nearly $1/2$ and the recombination process creates all other chromosomal types segregating with small total frequency. Which particular pair of principal complementary chromosomes are established depends on the initial population composition.

The assertion of part (i) can be extended to encompass a class of symmetric models with p alleles at each locus. To this end we concentrate on the special symmetric marginal fitness matrix of order $p \times p$ at locus i of the form

$$M_i = \left\| \begin{matrix} \alpha_i & 1 & \cdots & 1 \\ 1 & \alpha_i & \cdots & \\ \vdots & \vdots & \ddots & \vdots \\ 1 & 1 & \cdots & \alpha_i \end{matrix} \right\|, \tag{6.25}$$

where α_i occurs on the diagonal and 1 off the diagonal.

We stipulate $0 < \alpha_k < 1$ assuring that M_k manifests *strong total heterosis* in the sense of Definition 3.3.

Let the alleles possible at locus k be $A_1^{(k)}, A_2^{(k)}, \ldots, A_p^{(k)}$ which can be referred to by the numbers $1, 2, \ldots, p$.

The fitness matrix for the n-locus non-epistasis multiplicative system having marginal fitness matrix M_k of (6.25) at locus k is

$$M = M_1 \otimes M_2 \otimes \cdots \otimes M_n. \tag{6.26}$$

A chromosomal type available can now be described by a vector

$$C = (a^{(1)}, a^{(2)}, \ldots, a^{(n)}), \tag{6.27}$$

where each $a^{(i)}$ takes one of the values $1, 2, \ldots, p$ identifying which allele appears at that locus.

The totality of all chromosomal forms is denoted by $\mathcal{C}$. *A collection of chromosomal types* $C_i = (a_i^{(1)}, a_i^{(2)}, \ldots, a_i^{(n)})$, $i = 1, \ldots, p$, *is said to constitute a complementary set if* $\{a_i^{(k)}\}_{i=1}^p$ *for each* k *coincides in some arrangement with* $1, 2, \ldots, p$. In other words, each allele occurs exactly once at each locus among the array $C_1, C_2, \ldots, C_p$. Obviously a complementary set necessarily involves precisely p chromosomal types. To illustrate, for $n = 3$, $p = 3$,

$$(A_1^{(1)}, A_1^{(2)}, A_1^{(3)}), \quad (A_2^{(1)}, A_2^{(2)}, A_2^{(3)}), \quad (A_3^{(1)}, A_3^{(2)}, A_3^{(3)}),$$

and

$$(A_1^{(1)}, A_2^{(2)}, A_2^{(3)}), \quad (A_2^{(1)}, A_3^{(2)}, A_1^{(3)}), \quad (A_3^{(1)}, A_1^{(2)}, A_3^{(3)}),$$

constitute two sets of complementary chromosomes.

The next result generalizes part (i) of Theorem 6.3.

THEOREM 6.4: *Consider the one-locus selection regime* (6.26) *with marginal fitness matrices* (6.25) *having each* $0 < \alpha_i < 1$. *Let* $\Gamma = \{C_1, C_2, \ldots, C_p\}$ *consist of a set of complementary chromosomes. The frequency vector* $x(\Gamma)$ *defined on* $\mathcal{C}$ (*the totality of chromosomal types*) *which assigns probability* $1/p$ *to each element of* Γ *determines a stable equilibrium. Thus,* $x(\Gamma)$ *concentrates all its probability symmetrically on p complementary chromosomal types. There are* $(p!)^{n-1}$ *distinct sets of complementary chromosomes.*

The analog of part (ii) of Theorem 6.3 undoubtedly persists for Theorem 6.4.

It is important to emphasize that the high complementary is tied to the immense allelic symmetry endowed to the marginal fitness matrices comprising the selection regime (6.26). Where such symmetry is not present, then the stable equilibria remain boundary equilibria, but generally significantly more chromosomal types are represented as part of a stable configuration.

Additive selection model with *n* loci. Let M_i be the marginal fitness matrix corresponding to the ith locus. Let E_i be an $n_i \times n_i$ matrix whose elements *are all unity*. The fitness matrix for an n-locus trait with selection accruing additively over loci has the form

$$S = M_1 \otimes E_2 \otimes \cdots \otimes E_n + E_1 \otimes M_2 \otimes E_3 \otimes \cdots \otimes E_n$$
$$+ \cdots + E_1 \otimes \cdots \otimes E_{n-1} \otimes M_n = \sum_{k=1}^{n} S_k. \tag{6.28}$$

As in the multiplicative case, we have

THEOREM 6.5: *If* $\hat{x}_k$ *is an equilibrium for the kth locus fitness matrix* M_k, *then* $\hat{z} = \Pi_{k=1}^{n} \otimes \hat{x}_k$ *is a polymorphism for the transformation equation* (6.13) *associated with the additive fitness regime* (6.28).

The conditions for stability of $\hat{z}$ contrast sharply with the requirements in the corresponding multiplicative selection model.

THEOREM 6.6: *Suppose each of the single-locus polymorphism $\hat{x}_k$ (at locus k) is stable. Assume positive recombination parameters $R(\varepsilon) > 0$ for all ε. Then the polymorphism*

$$\hat{x} = \prod_{i=1}^{n} \otimes \hat{x}_i \tag{6.29}$$

of the additive fitness regime S is globally stable (with respect to all states having all chromosomal types initially represented).

For multiplicative non-epistasis, (6.29) can only be stable for sufficiently large recombination (loose linkage) while in the additive case overdominance at each separate locus already guarantees that the H. W. polymorphism is uniquely stable, for any degree of positive recombination.

7. ASPECTS OF MULTI-LOCUS PHENOTYPIC FITNESS ASSOCIATIONS

The underlying concept is that genotypes induce phenotype values by some prescribed association and fitness expression is a function of its phenotype value. We illustrate first with some two-locus two-allele examples. The first involves three phenotypes P_0, P_1, P_2, each determined by the cumulative heterozygosity over loci, viz.,

$$\overset{P_0}{\boxed{\frac{AB}{AB},\ \frac{Ab}{Ab},\ \frac{aB}{aB},\ \frac{ab}{ab}}} \qquad \overset{P_1}{\boxed{\frac{AB}{Ab},\ \frac{AB}{aB},\ \frac{Ab}{ab},\ \frac{aB}{ab}}} \qquad \overset{P_2}{\boxed{\frac{AB}{ab},\ \frac{Ab}{aB}}}. \tag{7.1}$$

Thus, all genotypes of phenotype class P_0 are double homozygotes, those of class P_1 single heterozygotes and those of class P_2 the double heterozygotes. Selection expression determined by that phenotype value means that the relative fitness value of all genotypes in P_0 coincide, those genotypes of phenotype corresponding to P_1 carry an identical fitness value and finally the genotypes of class P_2 have a single relative fitness value. The fitness array (cf. (5.5)) in this formulation takes the specific form

$$\begin{array}{c} \\ AB \\ Ab \\ aB \\ ab \end{array} \begin{array}{c} \begin{array}{cccc} AB & Ab & aB & ab \end{array} \\ \left\| \begin{array}{cccc} \alpha & \beta & \beta & \gamma \\ \beta & \alpha & \gamma & \beta \\ \beta & \gamma & \alpha & \beta \\ \gamma & \beta & \beta & \gamma \end{array} \right\| \end{array} = M. \tag{7.2}$$

This model is commonly referred to as the Lewontin-Kojima [1960] symmetric viability regime. Their perspective apparently was to regard the array (7.2) as reflecting symmetry with respect to labeling of the alleles. Other selection models in this vein were singled out by Wright [1952], Kimura [1956], Bodmer and Felsenstein [1967], and others; for a thorough analytic treatment, see Karlin and Feldman [1970b].

An important observation pertinent to the Lewontin-Kojima fitness matrix is that it admits the elegant Kronecker product representation (cf. (5.15))

$$M = \alpha I \otimes I + \beta[J \otimes I + I \otimes J] + \gamma J \otimes J \tag{7.3}$$

$$\text{where} \quad \boldsymbol{I} = \text{identity matrix and } J = \left\| \begin{matrix} 0 & 1 \\ 1 & 0 \end{matrix} \right\|.$$

This model and extensions of it are readily amenable to complete analysis owing to its simple structure inherent to (7.3).

Another important phenotypic classification carries five distinguishable phenotypes determined by additive allelic effects at the two loci where A, B confer allelic value one and a, b value zero. Accordingly, the 10 possible genotypes divide into 5 phenotype classes as indicated.

P_0	P_1	P_2	P_3	P_4
$\frac{ab}{ab}$	$\frac{aB}{ab}, \frac{Ab}{ab}$	$\frac{aB}{aB}, \frac{AB}{ab}, \frac{Ab}{aB}, \frac{Ab}{Ab}$	$\frac{AB}{aB}, \frac{AB}{Ab}$	$\frac{AB}{AB}$

The last phenotypic association has been used in the study of some phenomena of polygenic and quantitative inheritance; see Karlin and Carmelli [1975a].

Phenotypic model of additive heterozygosity effects involving *n* loci. An extended version of (7.3) to n loci has the following formulation:

Consider an n-loci trait involving two possible alleles at each of the different loci. Suppose the relative fitness of a genotype is determined by the number of heterozygous loci. Accordingly, there are $n + 1$ quantities γ_i, $i = 0, 1, 2, \ldots, n$, with the following interpretation:

$$\gamma_i = \text{the relative fitness of a genotype if } i \text{ among its loci are heterozygous.} \tag{7.4}$$

For $n = 3$ involving alleles $\{A, a\}$, $\{B, b\}$, $\{C, c\}$ at the respective loci the fitness matrix takes the form

$$\begin{array}{c} \\ ABC \\ ABc \\ AbC \\ Abc \\ aBC \\ aBc \\ abC \\ abc \end{array} \begin{array}{cccccccc} ABC & ABc & AbC & Abc & aBC & aBc & abC & abc \\ \end{array}$$

$$\begin{array}{c|cccccccc} & ABC & ABc & AbC & Abc & aBC & aBc & abC & abc \\ \hline ABC & \gamma_0 & \gamma_1 & \gamma_1 & \gamma_2 & \gamma_1 & \gamma_2 & \gamma_2 & \gamma_3 \\ ABc & \gamma_1 & \gamma_0 & \gamma_2 & \gamma_1 & \gamma_2 & \gamma_1 & \gamma_3 & \gamma_2 \\ AbC & \gamma_1 & \gamma_2 & \gamma_0 & \gamma_1 & \gamma_2 & \gamma_3 & \gamma_1 & \gamma_2 \\ Abc & \gamma_2 & \gamma_1 & \gamma_1 & \gamma_0 & \gamma_3 & \gamma_2 & \gamma_2 & \gamma_1 \\ aBC & \gamma_1 & \gamma_2 & \gamma_2 & \gamma_3 & \gamma_0 & \gamma_1 & \gamma_1 & \gamma_2 \\ aBc & \gamma_2 & \gamma_1 & \gamma_3 & \gamma_2 & \gamma_1 & \gamma_0 & \gamma_2 & \gamma_1 \\ abC & \gamma_2 & \gamma_3 & \gamma_1 & \gamma_2 & \gamma_1 & \gamma_2 & \gamma_0 & \gamma_1 \\ abc & \gamma_3 & \gamma_2 & \gamma_2 & \gamma_1 & \gamma_2 & \gamma_1 & \gamma_1 & \gamma_0 \end{array} = M^{(3)} \tag{7.5}$$

Some facets of the 3 loci selection model above with $\gamma_3 = 1 > \max(\gamma_0, \gamma_1, \gamma_2)$ were investigated by Feldman, Franklin and Thomson [1974]. They refer to this case as the completely symmetric three-locus model. Our emphasis is that the model of (7.5) naturally fits the framework of phenotypic fitness expression determined by cumulative heterozygosity of a genotype. Based on this perspective a number of extensions are readily forthcoming.

The matrix $M^{(3)}$ possesses a natural representation as a sum of Kronecker products of simple form

$$M^{(3)} = \gamma_0 I \otimes I \otimes I$$
$$+\gamma_1[J \otimes I \otimes I + I \otimes J \otimes I + I \otimes I \otimes J]$$
$$+\gamma_2[J \otimes J \otimes I + J \otimes I \otimes J + I \otimes J \otimes J] + \gamma_3 J \otimes J \otimes J,$$

$$\text{where} \quad I = \begin{Vmatrix} 1 & 0 \\ 0 & 1 \end{Vmatrix} \quad \text{and} \quad J = \begin{Vmatrix} 0 & 1 \\ 1 & 0 \end{Vmatrix}. \tag{7.6}$$

We now generalize the class of fitness matrices to the n-locus model corresponding to the phenotypic selection relationship prescribed in (7.4). To this end we introduce the following notation:

$$P_k^n(J) = \Sigma J^{\eta_1} \otimes J^{\eta_2} \otimes \cdots \otimes J^{\eta_n}, \ (J^0 = I), \tag{7.7}$$

where the sum extends over all specifications of n-tuples $\eta = (\eta_1, \eta_2, \ldots, \eta_n)$ with $\eta_i = 0$ or 1 and obeying $\Sigma_{i=1}^n \eta_i = k$. We emphasize that (7.7) is a sum of Kronecker products of n factors involving just the component matrices $I = \begin{Vmatrix} 1 & 0 \\ 0 & 1 \end{Vmatrix}$ and $J = \begin{Vmatrix} 0 & 1 \\ 1 & 0 \end{Vmatrix}$.

The matrix M of the phenotypic selection model corresponding to (7.4) can be represented as

$$M^{(n)} = \sum_{k=0}^{n} \gamma_k P_k^n(J). \tag{7.8}$$

We can write a complete system of eigenvalues for the phenotypic symmetric fitness matrix (7.8). These are

$$\lambda_l = \sum_{j=0}^{n} \gamma_j \left[\sum_{i=0}^{j} (-1)^i \binom{l}{i} \binom{n-l}{j-i} \right], \quad l = 0, 1, \ldots, n, \tag{7.9}$$

λ_l occurring with multiplicity $\binom{n}{l}$. The sign of the eigenvalues plays a key role in classifying and characterizing the equilibria of the transformation equation, connecting gamete frequencies over successive generations.

A more general n-locus phenotypic model with two alleles at each locus. Since two alleles are involved at each locus (say A_i and a_i at locus i), a chromosomal type can be discerned by an n-tuple

$$\xi = (\xi_1, \xi_2, \ldots, \xi_n), \tag{7.10}$$

where each $\xi_i = 0$ or 1 (1 signifying the occurrence of A_i and 0 stands for a_i). Henceforth, we arrange the chromosomal types in lexicographic order. Thus, for $n = 3$, the ordering is enumerated to be

$$(1, 1, 1), \quad (1, 1, 0), \quad (1, 0, 1), \quad (1, 0, 0), \quad (0, 1, 1), \quad (0, 1, 0), \quad (0, 0, 1), \quad (0, 0, 0).$$

A genotype is composed of two chromosomes (say ξ and ξ') and will be put in the notation

$$g(\xi, \xi') = \xi/\xi'. \tag{7.11}$$

Let L be a subset of indices from $(1, 2, \ldots, n)$. For each L (including the empty set), a positive number

$$\gamma(L), \quad L \text{ traversing the subsets of } \{1, 2, \ldots, n\}, \tag{7.12}$$

is prescribed endowed with an interpretation as a fitness value as described immediately below.

The fitness of a genotype $g = g(\xi, \xi')$ composed of chromosomes ξ and ξ' is determined as follows: We construct for g the set of indices

$$L(\xi, \xi') = \{ i \mid \xi_i = \xi_i' \}$$

embracing the loci indices where g is homozygous. Now define the fitness value of g to be

$$m(g(\xi, \xi')) = \gamma(L(\xi, \xi')). \tag{7.13}$$

Observe that the fitness value of a genotype is a function only of the positions where g is homozygous but not of the respective alleles appearing at these loci.

In order to secure a compact succinct representation of the global fitness matrix we single out the collection of 2^n matrices (cf. (7.7))

$$P(\eta) = J^{\eta_1} \otimes J^{\eta_2} \otimes \cdots \otimes J^{\eta_n} \tag{7.14}$$

determined for each $\eta = (\eta_1, \eta_2, \ldots, \eta_n)$, $\eta_i = 0$ or 1, subject to the convention

$$J^0 = I \quad \text{and} \quad J^1 = J = \begin{Vmatrix} 0 & 1 \\ 1 & 0 \end{Vmatrix}.$$

The fitness matrix Γ, keeping the lexicographic ordering, admits the representation

$$\Gamma = \sum_{\eta} \gamma(L(\eta)) P(\eta) \tag{7.15}$$

where $L(\eta) = \{i | \eta_i = 0\}$.

Making the appropriate identification of $\gamma(L)$ we readily find that the above representation formula subsumes the model (7.8).

It is not difficult to work out the eigenvalues of (7.15) explicitly. Thus with each prescription of $\varepsilon = (\varepsilon_1, \ldots, \varepsilon_n)$, $\varepsilon_i = 0$ or 1, we get an eigenvalue

$$\lambda(\varepsilon) = \sum_{\eta} \gamma(\eta) \Big(\prod_{i=1}^{n} (-1)^{\eta_i \varepsilon_i} \Big) \text{ where } \gamma(\eta) = \gamma(L(\eta)).$$

A set of *n*-loci viability arrays with common symmetric eigenstructure. A general pattern of viability matrices subsuming (7.8) and (5.17) has the structure of sums of Kronecker products of special form

$$M = \sum_{k=1}^{m} c_k Q_{\alpha^{(k)}}, \quad (c_k > 0, k = 0, 1, \ldots, m), \tag{7.16}$$

where

$$Q_{\alpha^{(k)}} = U_{\alpha_1^{(k)}} \otimes U_{\alpha_2^{(k)}} \otimes \cdots \otimes U_{\alpha_n^{(k)}} = \prod_{j=1}^{n} \otimes \begin{Vmatrix} \alpha_j^{(k)} & 1 \\ 1 & \alpha_j^{(k)} \end{Vmatrix} \tag{7.17}$$

and $\alpha_j^{(k)}$ are non-negative constants.

The phenotypic models (7.8) and (7.15) are manifestly important subclasses of (7.16).

Each term of (7.16) generates a multiplicative selection model. The transformation relations connecting gamete frequencies in successive generations achieve an elegant expression:

For $Q = \Pi_{i=1}^{n} \otimes U_{\alpha_i}$, we have by (6.14)

$$x' = \sum_{\varepsilon} R(\varepsilon)\left(U_{\alpha_1}^{\varepsilon_1} \otimes U_{\alpha_2}^{\varepsilon_2} \otimes \cdots \otimes U_{\alpha_n}^{\varepsilon_n}\right)x \circ$$

$$\frac{\left(U_{\alpha_1}^{1-\varepsilon_1} \otimes \cdots \otimes U_{\alpha_n}^{1-\varepsilon_n}\right)x}{(x, Qx)} = T_Q x, \tag{7.18}$$

where T_Q is a compact notation for the transformation herein.

The total recurrence relations of gamete frequencies in two successive generations for the viability matrix M (of (7.16)) becomes (abbreviating $Q^{(k)}$ for $Q_{\alpha^{(k)}}$ and T_k for $T_{Q^{(k)}}$)

$$x' = \frac{\sum_{k=1}^{m} c_k(x, Q^{(k)}x)T_k x}{\sum_{k=1}^{m} c_k(x, Q^{(k)}x)}. \tag{7.19}$$

The vector

$$u^* = \frac{1}{2^n}(u \otimes u \otimes \cdots \otimes u), \quad u = (1, 1) \tag{7.20}$$

is a common Hardy-Weinberg (H. W.) equilibrium (meaning that it is a Kronecker product and accordingly its higher order measures of association are zero) for each transformation T_k and therefore also an equilibrium of the total transformation relations (7.19). At the close of this section the complete criteria guaranteeing stability of u^* will be delineated.

The component marginal fitness matrices

$$U_\alpha = \begin{Vmatrix} \alpha & 1 \\ 1 & \alpha \end{Vmatrix}, \quad 0 \leqslant \alpha \leqslant \infty,$$

occurring in (7.17) (interpret the specification $\alpha = \infty$ as the identity matrix $U_\infty = I$) share the eigenvectors $u = (1, 1)$ and $e = (1, -1)$ with corresponding eigenvalues $\alpha + 1$ and $\alpha - 1$, respectively. Note that the J that features so prominently in the cases of (7.8) and (7.15) is merely $J = U_0$.

A natural basis of eigenvectors for the Kronecker product matrix (7.17) are the 2^n vectors

$$v(\delta) = \prod_{i=1}^{n} \otimes e^{\delta_i} = e^{\delta_1} \otimes e^{\delta_2} \otimes \cdots \otimes e^{\delta_n} \tag{7.21}$$

generated by all possible specifications $\delta = (\delta_1, \delta_2, \ldots, \delta_n)$, $\delta_i = 1$ or 0, subject to the convention that $e^0 = u$ and $e^1 = e$. This system of vectors is mutually orthogonal as each Q is a symmetric matrix (or this fact can be checked by direct calculations).

The collection of 2^n vectors (7.21) each of 2^n coordinates, are endowed with two basic properties:

(i) The family of vectors $\{v(\delta)\}_\delta$ which constitute an orthogonal base in 2^n dimensional Euclidean space are simultaneously eigenvectors for each summand of (7.16). (7.22)

(ii) The set $\{v(\delta)\}$ form a *group* with respect to the "$\circ$" operation (Schur product, see Appendix A). (7.23)

Inspection of the transformation equation (7.18) reveals that the Schur product operation occurs in an intrinsic way. Property (ii) facilitates the task of analyzing the equilibrium structure of the models induced by the fitness matrix (7.16) having a general set of recombination frequencies; consult Karlin [1979] for more on this.

Local stability of the central equilibrium $u^* = \frac{1}{2^n}(u \otimes u \otimes \cdots \otimes u)$. An adaptation of simple pertubation methods produces the gradient matrix (local linear approximation at u^*)

$$\zeta' = \frac{2 \sum_{k=1}^{m} c_k (Q^{(k)}u^*, u^*) \sum_{\delta} R(\delta) \prod_{i=1}^{n} \otimes \tilde{U}_{\alpha_i(k)}^{\delta_i} \zeta}{\sum_{k=1}^{m} c_k (Q^{(k)}u^*, u^*)} = L\zeta, \quad (7.24)$$

where

$$\tilde{U}_\alpha = \frac{u \circ U_\alpha}{2(U_\alpha u, u)} = \frac{1}{\alpha + 1} u \circ U_\alpha. \quad (7.25)$$

The relevant domain $\mathfrak{D}$ of the linear mapping (7.24) is the orthogonal complement to u^*. This space is spanned by the system of $2^n - 1$ vectors

$$\{v(\delta)\}, \delta \neq (1, 1, \ldots, 1). \quad (7.26)$$

Note that the eigenvalues of $\tilde{U}_\alpha$ are 1 and $(\alpha - 1)/(\alpha + 1)$ with corresponding eigenvectors u and e, respectively.

The collection in (7.26) constitutes all relevant eigenvectors of the mapping L in (7.24). Substituting each $v(\delta)$ into (7.24), we can ascertain the eigenvalues of L and thereby achieve the result:

u^* is locally stable if and only if the inequalities

$$2\left| \frac{\sum_{k=1}^{m} c_k \prod_{i=1}^{n} [\alpha_i^{(k)} + 1] \sum_{\delta} R(\delta) \lambda_{k,1}^{\delta_1} \lambda_{k,2}^{\delta_2} \cdots \lambda_{k,n}^{\delta_n}}{\sum_{k=1}^{m} c_k \prod_{i=1}^{n} [\alpha_i^{(k)} + 1]} \right| < 1 \quad (7.27)$$

hold for all choices of $\lambda_{k,j}$, where for a given j and all k, either $\lambda_{k,j} = 1$ or $\frac{\alpha_j^{(k)} - 1}{\alpha_j^{(k)} + 1}$, with the single exception that the choice $\lambda_{k,j} \equiv 1$ for all j is precluded. Some further analysis establishes the important result:

THEOREM 7.1: *Consider the multi-locus selection model with fitness matrix of the structure* (7.16). *If the central polymorphism* u^* (*see* (7.20)) *is stable in the case of absolute linkage* (*no recombination*), *then* u^* *is stable for any set of recombination frequencies.*

We conjecture that the above result is a general phenomenon without any restriction on the fitness matrix. Its statement for an arbitrary multi-locus multi-allele selection recombination system is as follows: *Suppose there exists a stable polymorphic equilibrium in the presence of no recombination, then there exists a unique stable polymorphism for any positive recombination.* Naturally, the polymorphic state generally depends continuously on the recombination frequencies. In these circumstances, the selection balance is an overwhelming influence maintaining a polymorphism while the recombination mechanism exercises minor effects (mere displacement of the equilibrium but retaining it approximately centrally located).

If in the absence of recombination where the selection forces cannot maintain a polymorphism, then the recombination process becomes an important force and the polymorphisms that do arise usually involve a partial set of chromosomal types each represented with moderate frequency while the remaining gametes occur with minute frequency sustained by virtue of recombination. Moreover, in these situations there is usually a *multiplicity* of polymorphisms. The finding of Theorem 7.1 suggests a dichotomy in the nature of stable polymorphic states.

Indeed, there are two types of polymorphisms occurring in multi-locus selection models:

(i) A balance of the relative genotypic selection forces pervades the evolutionary process establishing a stable polymorphic state while the recombination mechanism exercises minor or trivial effects entailing at most a small displacement in the location of the equilibrium points.

(ii) Selection forces alone (without recombination) would lead to the elimination of certain chromosomal types but in the presence of some recombination rate the full complement of chromosomal types are created and sustained.

The characteristics of the polymorphic outcomes for the circumstances of (i) and (ii) are markedly different. For the situation of (i), usually a single, rather central, globally stable polymorphism is manifested for each set of recombination rates. Equivalently, the induced multi-allele one-locus model corresponding to no recombination but subject to balancing selection forces essentially determines the equilibrium configuration which usually persists for all values of the recombination parameters. In the situation (ii), one or more stable polymorphisms are created (mostly present for tight linkage) with the property that the chromosomal types divide into two groups such that the frequencies in one group occur in trace amounts while each of the chromosomal types in the other group are represented with more substantial frequency. Such polymorphisms are accordingly located near the boundary of the frequency simplex.

8. MEAN FITNESS OVERDOMINANCE AT EACH LOCUS INDUCED FROM A STABLE MULTI-LOCUS POLYMORPHISM

We conclude this article with a general analytic finding of some biological interest. Consider the fitness regime of (5.5). For a given set of gamete frequencies $x = (x_1, x_2, x_3, x_4)$, we determine the induced mean fitness values of the various genotypes at the first locus by the explicit formulas:

$$\textit{Induced Fitness of } AA = W_{AA}(x) =$$

$$\frac{x_1^2\left(\text{fitness } \dfrac{AB}{AB}\right) + 2x_1x_2\left(\text{fitness } \dfrac{AB}{Ab}\right) + x_2^2\left(\text{fitness } \dfrac{Ab}{Ab}\right)}{x_1^2 + 2x_1x_2 + x_2^2}$$

$$= \frac{x_1^2w_{11} + 2x_1x_2w_{12} + x_2^2w_{22}}{x^2 + 2x_1x_2 + x_2^2}, \tag{8.1}$$

$$W_{Aa}(x) = \frac{x_1x_3w_{13} + x_1x_4w_{14} + x_2x_3w_{23} + x_2x_4w_{24}}{x_1x_3 + x_1x_4 + x_2x_3 + x_2x_4}$$

$$W_{aa}(x) = \frac{x_3^2w_{33} + 2x_3x_4w_{34} + x_4^2w_{44}}{x_3^2 + 2x_3x_4 + x_4^2}.$$

The relevance and interest in such induced fitness values apparently was first underscored by Bodmer and Felsenstein [1967] and later referenced by Franklin and Lewontin [1970].

Let $\hat{x} = (\hat{x}_1, \hat{x}_2, \hat{x}_3, \hat{x}_4)$ be a stable polymorphic equilibrium. We calculate the mean induced fitness values at locus $A - a$ yielding

$$W_{AA}(\hat{x}),\ W_{Aa}(\hat{x}),\ W_{aa}(\hat{x}). \tag{8.2}$$

A question of practical utility concerns whether the array (8.2) displays heterozygote advantage. Equivalently, does a stable *polymorphic* multi-locus equilibrium necessarily engender mean fitness overdominance at each separate locus?

We can establish from Theorem 3.1 the following general result:

THEOREM 8.1: *Consider any multi-locus (two alleles at each locus) viability model with small recombination frequencies. Let x^* be a stable polymorphism and evaluate the mean fitness values at the $A - a$ locus, viz., $W_{AA}(x^*)$, $W_{Aa}(x^*)$, $W_{aa}(x^*)$. Then*

$$W_{Aa}(x^*) > \max(W_{AA}(x^*), W_{aa}(x^*)) \tag{8.3}$$

indicating that the induced mean fitness parameters at the $A - a$ locus manifests heterozygote advantage. (Of course, the same property persists for each separate locus.)

We do not know whether the conclusion of (8.3) applies for loose linkage with general selection coefficients. (Numerical studies suggest that it is correct.) *Definitely, for the classical models (additive, multiplicative and symmetric viability regimes; see Sections* 6 *and* 7) *the tenet of* (8.3) *prevails for any stable polymorphism and recombination rate.*

The following interpretation is a valid corollary of the above result: If the observations at a locus significantly preclude an explanation based on overdominance, then segregation at the locus in question cannot be ascribed to a supergene complex maintaining polymorphism solely by a complex of selection forces.

There is a multi-locus version of (8.3). Consider a general selection regime for a collection of l loci and involving an arbitrary set of small recombination frequencies, and let x^* be a stable polymorphism. *We form the induced fitness values* $W_{ij}(x^*)$, $1 \leqslant i, j \leqslant k$, *for* x^* *with respect to a specified* k *loci* ($k < l$). *Then the one-locus* k*-allele selection model with fitness coefficients* $W_{ij}(x^*)$ *entails a unique globally stable polymorphism.*

BIBLIOGRAPHY

W. F. Bodmer, "Differential fertility in population genetic models," *Genetics,* **51** (1965), 411–424.

W. F. Bodmer and J. Felsenstein, "Linkage and selection: Theoretical analysis of the deterministic two-locus random mating model," *Genetics,* **57** (1967), 237–265.

C. Cannings, "Equilibrium, convergence and stability at a sex linked locus under natural selection," *Genetics,* **56** (1967), 613–617.

———, "Unisexual selection at an autosomal locus," *Genetics,* **62** (1969a), 225–229.

———, "The study of multiallelic genetic systems by matrix methods," *Genet. Res.,* **14** (1969b), 167–183.

F. B. Christiansen and M. W. Feldman, "Subdivided populations: a review of the one and two locus deterministic theory," *Theor. Pop. Biol.,* **7** (1974), 13–38.

———, "Selection in complex genetic systems IV: Multiple alleles and interactions between two loci," *J. Math. Biol.,* **2** (1975), 179–205.

J. F. Crow and M. Kimura, *An Introduction to Population Genetics Theory,* Harper and Row, New York, 1970.

I. Eshel and M. W. Feldman, "On the evolutionary effect of recombination," *Theor. Pop Biol.,* **1** (1970), 88–100.

W. J. Ewens, "A genetic model having complex linkage behavior," *Theor. Appl. Genet.,* **1** (1968), 140–143.

———, "Mean fitness increases when fitnesses are additive," *Nature,* **221** (1969a), 1076.

———, *Population Genetics,* Methuen Press, London, 1969b.

M. W. Feldman, "Equilibrium studies of two-locus haploid populations with recombination," *Theor. Pop. Biol.,* **2** (1971), 299–318.

———, "Selection for linkage modification: I. Random mating populations," *Theor. Pop. Biol.,* **3** (1972), 324–346.

M. W. Feldman, I. R. Franklin, and G. Thomson, "Selection in complex genetic systems: I. The symmetric equilibrium of the three locus symmetric viability model," *Genetics,* **76** (1974), 135–162.

M. W. Feldman and J. Krakauer, "Genetic modification in modifier polymorphisms," *Conf. Proc. Population Genetics and Ecology* (S. Karlin and E. Nevo, eds.), Academic Press, New York, 1976, 547–584.

M. W. Feldman, R. C. Lewontin, I. R. Franklin, and F. B. Christiansen, "Selection in complex genetic systems III. An effect of allele multiplicity with two loci," *Genetics,* **79** (1975), 333–347.

I. Franklin and R. C. Lewontin, "Is the gene the unit of selection?," *Genetics,* **65** (1970), 701–734.

J. H. Gillespie and C. Langley, "Multilocus behavior in random environments: I. Random Levene models," *Genetics,* **82** (1976), 123–137.

W. G. Hill, "Disequilibrium among several linked neutral genes in finite populations: II. Variances and covariances of disequilibrium," *Theor. Pop. Biol.,* **6** (1974), 181–198.

———, "Non-random association of neutral linked genes in finite populations," *Population Genetics and Ecology* (S. Karlin and E. Nevo, eds.), Academic Press, New York, 1976, 339–376.

S. Karlin, *Mathematical Methods and Theory in Games, Programming and Economics,* vols. I and II, Addison-Wesley, Reading, 1959.

———, "Equilibrium behavior of population genetic models with non-random mating," *J. Appl. Prob.,* **5** (1968), 231–313, 487–566.

———, "Some mathematical models of population genetics, *Amer. Math. Monthly,* **79** (1972), 699–739.

———, "Sex and infinity: A mathematical analysis of the advantage and disadvantage of genetic recombinations," *Population Dynamics* (M. Bartlett and R. Hiorns, eds.), Academic Press, New York, 1973.

———, "General two-locus selection models: Some objectives, results and interpretations," *Theor. Pop. Biol.,* **7** (1975), 364–398.

———, "Population subdivision and selection migration interactions," *Population Genetics and Ecology* (S. Karlin and E. Nevo, eds.), Academic Press, New York, 1976, 617–659.

———, "Studies of selection in multilocus systems. II. Selection depending on aggregate genotypic heterozygosity,"*Proc. of Conf. on Quantitive Genetics* (E. Pollak and O. Kempthorne, eds.), Iowa State University Press, 1977.

———, "Studies of selection in multilocus systems. III. Extended symmetric heterotic selection regimes," (1978a), to appear.

———, "Some gaussian models of polygenic and quantitative genetics," *Theor. Pop. Biol.,* (1978b), to appear.

———, *Theoretical Population Genetics,* Academic Press, New York, 1979, to appear.

S. Karlin and D. Carmelli, "Some population genetic models combining artificial and natural selection pressures," *Proc. Nat. Acad. Sci.,* **71** (1975a), 4724–4731.

———, "Numerical studies on two-loci selection models with general viabilities," *Theor. Pop. Biol.,* (1975b), 399–421.

S. Karlin and N. Richter-Dyn, "Some theoretical analyses of migration selection interaction in a cline: A generalized two-range environment," *Conf. Proc. Population Genetics and Ecology* (S. Karlin and E. Nevo, eds.), Academic Press, New York, 1976, 659–707.

S. Karlin and S. Farkash, "Equilibrium structure of assortative mating systems with multiple loci," 1978 (to appear).

S. Karlin and M. W. Feldman, "Convergence to equilibrium of the two-locus additive viability model," *J. Appl. Prob.*, **7** (1970a), 262–271.

———, "Linkage and selection: Two-locus symmetric viability model," *Theor. Pop. Biol.*, **1** (1970b), 39–71.

S. Karlin and U. Lieberman, "Random temporal variation in selection intensities: Case of large population size," *Theor. Pop. Biol.*, **6** (1974), 355–382.

———, "A phenotypic symmetric selection model for three loci, two alleles. The case of tight linkage," *Theor. Pop. Biol.*, **10** (1976).

———, "Analysis and representations of some nonepistatis selection models," 1977, to appear.

S. Karlin and J. L. McGregor, "Rates and probabilities of fixations for two-locus random mating finite populations without selection," *Genetics*, **58** (1968), 141–159.

———, "On mutation selection balance for two-locus haploid and diploid populations," *Theor. Pop. Biol.*, **2** (1971), 60–70.

———, "Polymorphisms for genetic and ecological systems with weak coupling," *Theor. Pop. Biol.*, **3** (1972a), 210–238.

———, "The evolutionary development of modifier genes," *Proc. Nat. Acad. Sci. USA*, **69** (1972b), 3611–3614.

———, "Towards a theory of the evolution of modifier genes," *Theor. Pop. Biol.*, **5** (1974), 59–103.

O. Kempthorne, *An Introduction to Genetic Statistics*, Wiley, New York, 1957.

J. F.C. Kingman, "A matrix inequality," *Quart. J. Math.*, **12** (1961a), 78–80.

———, "A mathematical problem in population genetics," *Proc. Cambridge Philos. Soc.*, **57** (1961b), 574–582.

M. Kimura, "A model of a genetic system which leads to closer linkage by natural selection," *Evolution*, **10** (1956), 278–281.

———, "On the change of population fitness by natural selection," *Heredity*, **12** (1958), 145–167.

M. Kimura and T. Ohta, *Theoretical Aspects of Population Genetics*, Princeton University Press, Princeton, 1971.

K. Kojima and R. C. Lewontin, "Evolutionary significance of linkage and epistasis," *Topics in Math. Genetics* (K. Kojima, ed.), Springer-Verlag, New York, 1970.

R. Lande, "The maintenance of genetic variability by mutation in a polygenic character with linked loci," *Genet. Res. Camb.*, **26** (1976), 221–235.

R. C. Lewontin and K. Kojima, "The evolutionary dynamics of complex polymorphism," *Evolution*, **14** (1960), 458–472.

R. C. Lewontin, *The Genetical Basis for Evolutionary Change*, Columbia University Press, New York, 1974.

W. H. Li and M. Nei, "Stable linkage disequilibrium without epistasis in subdivided populations," *Theor. Pop. Biol.*, **6** (1974), 173–183.

S. P. H. Mandel, "The stability of a multiple allelic system," *Heredity*, **13** (1959), 289–302.

J. Maynard Smith, "What use is sex?", *J. Theor. Biol.* **30**, 319–335.

P. A. P. Moran, "Balanced polymorphisms with unlinked loci," *Aust. J. Biol. Sci.*, **16** (1963), 1–5.

———, "On the theory of selection dependent on two loci," *Ann. Hum. Genet.*, **32** (1968), 183–190.

T. Nagylaki and J. F. Crow, "Continuous selective models," *Theor. Pop. Biol.*, **5** (1974), 257–283.

M. Nei, "Modification of linkage intensity by natural selection," *Genetics*, **57** (1967), 681–699.

T. Ohta and M. Kimura, "Linkage disequilibrium due to random genetic drift," *Genet. Res.*, **13** (1969), 47–55.

A. R. G. Owens, "A genetical system admitting of two distinct stable equilibria under natural selection," *Heredity*, **7** (1953), 97–102.

C. Z. Roux, "Genetic loads: I. Hardy-Weinberg equilibria in random mating populations," *Theor. Pop. Biol.*, **5** (1974), 393–416.

P. A. G. Scheuer and S. P. H. Mandel, "An inequality in population genetics," *Heredity*, **13**, (1959), 519–524.

C. Strobeck, "The three-locus model with multiplicative fitness values: The crystallization of the genome," *Population Genetics and Ecology* (S. Karlin and E. Nevo, eds.), Academic Press, New York, 1976, 781–791.

J. A. Sved, "The stability of linked systems of loci with a small population size," *Genetics*, **59** (1968), 543–564.

S. Wright, "The genetics of quantitative variability," *Quantitative Inheritance* (K. Mather, ed.), Her Majesty's Stationery Office, London, 1952, 5–41.

THE GEOGRAPHICAL STRUCTURE OF POPULATIONS*

Thomas Nagylaki†

1. INTRODUCTION

The maintenance of genetic variability is one of the central problems of population genetics and evolutionary theory. By comparing theoretical predictions based on various hypotheses to experimental data, one hopes to evaluate the relative importance of different mechanisms in determining the genetic structure and evolution of natural populations. Many of the schemes which have been proposed for preserving genetic diversity were listed recently in [1].

In this chapter, we shall discuss the amount and pattern of genetic variation in geographically structured populations. Taking

*Paper No. 1826 from the Genetics Laboratory, University of Wisconsin. Sponsored by the United States Army (Contract No. DA-31-124-ARO-D-462), the National Institutes of Health (Grant GM-15422), and the National Science Foundation (Grant MPS72-00381 A01).

†Present address: Department of Biophysics and Theoretical Biology, University of Chicago, Cummings Life Science Center, 920 East 58th Street, Chicago, Illinois 60637.

spatial distribution into account makes the treatment of genetic polymorphism richer in interest and more realistic for natural populations. In Section 2, we shall consider selectively neutral models with migration, mutation, and random genetic drift. In Section 3, we shall analyze the spatial variation in gene frequency under the combined action of migration and selection.

To achieve some coherence of biological and mathematical formulation, a few topics properly subsumed under the title of this article will be omitted. For the pioneering work of Wright on the random drift model with migration, the reader may refer to [2, Ch. 12]. For some theorems on multiniche models with migration and selection, applications to two niches, and a review of previous two-niche schemes, he may consult [3]. A very interesting new branching diffusion approach permitting the simultaneous analysis of selection and random drift is described in [4]. Refer to the "Note added in proof" at the end for other topics and more recent work.

Adequate definitions of the genetic terms occurring in this paper may be found in [5]. The concise and lucid exposition of basic genetics in [6] more than suffices for elementary background. The books of Crow and Kimura [7], Ewens [8], Kimura and Ohta [9], Malécot [10], Moran [11], and Wright [2] provide different perspectives on theoretical population genetics. Felsenstein and Taylor have compiled a bibliography of the field [12] with almost complete coverage through the autumn of 1973.

2. NEUTRAL GENES WITH RANDOM DRIFT

In this section, we shall study the relation between individuals as a function of time and their location. We confine ourselves to a single locus in the absence of selection, and assume every new mutant is of a type not preëxisting in the population. Diploid (i.e., with paired chromosomes) monoecious (= hermaphroditic) individuals are formed by the random fusion of proximate gametes. The habitat has n dimensions, and may be finite or infinite. Let $f(t, \mathbf{x}, \mathbf{y})$ be the probability that at time t two homologous genes, one at $\mathbf{x}$ and the other at $\mathbf{y}$, are the same allele. Following Maruyama, we shall call f the probability of allelism. The probability of nonallelism, $h(t, \mathbf{x}, \mathbf{y}) = 1 - f(t, \mathbf{x}, \mathbf{y})$, is clearly a measure of the genetic variability in the population, for if all genes have the same allelic state, then evidently $f(t, \mathbf{x}, \mathbf{y}) = 1$, and hence $h(t, \mathbf{x}, \mathbf{y}) = 0$. The probability that an individual at $\mathbf{x}$ is homozygous is $f(t, \mathbf{x}, \mathbf{x})$.

The problem may be posed in terms of two other closely related variables. The use of the probability of identity by descent [10, pp. 1–11, 64–84; 13] requires only a different initial condition from that for the probability of allelism. For instance, the population may be initially completely homogeneous, so that $f(0, \mathbf{x}, \mathbf{y}) = 1$, but all genes may, in principle, be nonidentical by descent. In the presence of mutation, the conceptual interpretation and experimental measurement in natural populations of the probability of allelism is much more direct than that of the probability of identity by descent. In a model in which all new mutants are of a novel type, the equilibrium values of the two probabilities are the same. For a diallelic model, it is sometimes more convenient to employ the correlation between gene frequencies at different locations [14, 15, 16]. To compensate for a longer formulation, the boundary conditions are more transparent if one works directly with gene frequencies, and at equilibrium strong stabilizing selection may be dealt with in the linear approximation [10, p. 65].

The stepping-stone model, in which generations are discrete and nonoverlapping, and panmictic (= randomly mating) colonies located at the points of an n-dimensional lattice exchange migrants, was proposed independently by Malécot [17] and Kimura [18]. We shall discuss a very general version of this model in Section 2.1. We pass to a continuous spatial distribution [10, pp. 64-84; 13] in Section 2.2, and treat populations continuous in space and time [10, pp. 64-84; 13] in Section 2.3.

2.1. The stepping-stone model. We suppose all alleles mutate to new types at rate u, and the number of diploid monoecious organisms in the colony at lattice point $\mathbf{x}$ at time t is $N(t, \mathbf{x})$. $N(t, \mathbf{x})$ varies deterministically. The probability that a randomly chosen individual at $\mathbf{y}$ at time $t + 1$ was at $\mathbf{x}$ at time t will be denoted by $m(t, \mathbf{x}, \mathbf{y})$. If N is independent of $\mathbf{x}$ and t, it is easy to see that m is the same as the probability density of final positions, given initial ones [10, p. 64]. For two genes in generation $t + 1$ to be identical in state, mutation must not occur, and they must be descended from either the same gene or identical genes. Thus, the basic equation of the model reads

$$f(t+1,\mathbf{x},\mathbf{y}) = (1-u)^2\Big\{\sum_{\mathbf{zw}} m(t,\mathbf{z},\mathbf{x})m(t,\mathbf{w},\mathbf{y})f(t,\mathbf{z},\mathbf{w})$$
$$+\sum_{\mathbf{z}}\left[2N(t,\mathbf{z})\right]^{-1}m(t,\mathbf{z},\mathbf{x})m(t,\mathbf{z},\mathbf{y})\left[1-f(t,\mathbf{z},\mathbf{z})\right]\Big\}. \quad (1)$$

The second sum in (1) corrects the term with $\mathbf{w}=\mathbf{z}$ in the first one for the possibility that the genes at $\mathbf{x}$ and $\mathbf{y}$ may come from the same colony at $\mathbf{z}$, in which case they may originate from the same gene, and then the appropriate probability of identity in generation t is unity, not $f(t,\mathbf{z},\mathbf{z})$. The probability of descent from the same gene, given descent from the same colony, is manifestly just $[2N(t,\mathbf{z})]^{-1}$, the reciprocal of the number of genes in the colony. Equation (1) presupposes that migration is independent of genotype, and no colony ever becomes empty.

Since the two genes are not distinguished, we may choose the symmetric initial condition $f(0,\mathbf{x},\mathbf{y})=f(0,\mathbf{y},\mathbf{x})$. It is easy to see from (1) that this relation will then hold for all subsequent times.

We proceed at once to the analysis of the simplest special case of (1). Let us assume $N(t,\mathbf{x})=N$, independent of t and $\mathbf{x}$, migration depends only on displacement, not on time and initial and final positions: $m(t,\mathbf{x},\mathbf{y})=m(\mathbf{y}-\mathbf{x})$ (to impede the proliferation of notation, we have not renamed the migration function), and the infinite lattice has no boundaries. We posit also that the initial probability of allelism is a function only of separation: $f(0,\mathbf{x},\mathbf{y})=f(0,\boldsymbol{\xi})$, where $\boldsymbol{\xi}=\mathbf{x}-\mathbf{y}$. The manipulations to follow imply that, as is intuitively obvious, this property is preserved as the population evolves. Extending the double sum in (1) to the second term by introducing the n-dimensional Kronecker delta $\delta_{\mathbf{zw}}$ and defining $\mathbf{X}=\mathbf{z}-\mathbf{w}$, we may rewrite (1) as

$$f(t+1,\boldsymbol{\xi}) = (1-u)^2\sum_{\mathbf{zX}} m(\mathbf{x}-\mathbf{z})m(\mathbf{x}-\boldsymbol{\xi}+\mathbf{X}-\mathbf{z})$$
$$\times\left\{f(t,\mathbf{X})+(2N)^{-1}\left[1-f(t,\mathbf{0})\right]\delta_{\mathbf{X0}}\right\}. \quad (2)$$

With the substitutions $\boldsymbol{\eta} = \mathbf{X} - \boldsymbol{\xi}$ and $\boldsymbol{\zeta} = \mathbf{x} - \mathbf{z}$, (2) becomes

$$f(t+1, \boldsymbol{\xi}) = (1-u)^2 \sum_{\boldsymbol{\eta}} r(\boldsymbol{\eta}) \left\{ f(t, \boldsymbol{\eta} + \boldsymbol{\xi}) + (2N)^{-1}\left[1 - f(t, \mathbf{0})\right] \delta_{\boldsymbol{\eta}+\boldsymbol{\xi}, \mathbf{0}} \right\}, \quad (3)$$

where

$$r(\boldsymbol{\eta}) = \sum_{\boldsymbol{\zeta}} m(\boldsymbol{\zeta}) m(\boldsymbol{\zeta} + \boldsymbol{\eta}) \quad (4)$$

is the probability that the separation between two alleles changes by $\boldsymbol{\eta}$ in one generation [13, 19]. Evenness is an important intuitive property of the probability density r. Indeed, (4) yields, with the aid of the substitution $\mathbf{Y} = \boldsymbol{\zeta} - \boldsymbol{\eta}$,

$$r(-\boldsymbol{\eta}) = \sum_{\boldsymbol{\zeta}} m(\boldsymbol{\zeta}) m(\boldsymbol{\zeta} - \boldsymbol{\eta}) = \sum_{\mathbf{Y}} m(\boldsymbol{\eta} + \mathbf{Y}) m(\mathbf{Y}) = r(\boldsymbol{\eta}). \quad (5)$$

Note that m need not be even for (5) to hold [20]. Employing (5) in (3), we deduce the equation satisfied by the probability of allelism:

$$f(t+1, \boldsymbol{\xi}) = (1-u)^2 \left\{ \sum_{\boldsymbol{\eta}} r(\boldsymbol{\eta}) f(t, \boldsymbol{\xi} - \boldsymbol{\eta}) + (2N)^{-1}\left[1 - f(t, \mathbf{0})\right] r(\boldsymbol{\xi}) \right\}. \quad (6)$$

The symmetry of $f(t, \mathbf{x}, \mathbf{y})$ mentioned in connection with (1) implies for (6) that $f(t, \boldsymbol{\xi})$ is even: $f(t, -\boldsymbol{\xi}) = f(t, \boldsymbol{\xi})$. This can also be proved directly from (5) and (6).

The rate and nature of the approach to equilibrium are obtained in [40], and some analyses are presented in the next two subsections. Here, we restrict ourselves to the equilibrium probability of allelism, $f_\infty(\boldsymbol{\xi})$. As suggested by the convolution structure of (6), we define [10, pp. 77–84; 15]

$$\varphi_\infty(\boldsymbol{\theta}) = \sum_{\boldsymbol{\xi}} f_\infty(\boldsymbol{\xi}) e^{-i\boldsymbol{\theta}\cdot\boldsymbol{\xi}}, \quad (7)$$

where the dot signifies the scalar product, and derive directly from the equilibrium form of (6) that the function $\varphi_\infty(\boldsymbol{\theta})$ is given by

$$\varphi_\infty(\boldsymbol{\theta}) = (1-u)^2 R(\boldsymbol{\theta})\left\{\varphi_\infty(\boldsymbol{\theta}) + (2N)^{-1}\left[1 - f_\infty(\mathbf{0})\right]\right\}. \quad (8)$$

In (8),

$$R(\boldsymbol{\theta}) = \sum_{\xi} r(\xi) e^{i\boldsymbol{\theta}\cdot\xi}. \quad (9)$$

Now (7) and (8) allow us to calculate the Fourier coefficients $f_\infty(\xi)$ of $\varphi_\infty(\boldsymbol{\theta})$:

$$f_\infty(\xi) = \frac{(1-u)^2\left[1 - f_\infty(\mathbf{0})\right]}{2N} \int_{-\pi}^{\pi} \frac{d^n\theta}{(2\pi)^n} \frac{e^{i\xi\cdot\boldsymbol{\theta}} R(\boldsymbol{\theta})}{1-(1-u)^2 R(\boldsymbol{\theta})}. \quad (10)$$

Of course, the notation means that θ_i is integrated from $-\pi$ to π for $i = 1, 2, \ldots, n$. Since $R(\boldsymbol{\theta})$ is known, and $f_\infty(\mathbf{0})$ may, in principle, be evaluated by setting $\xi = \mathbf{0}$ in (10), this completes the solution. In one and two dimensions, (10) may be approximated quite generally for $u \ll 1$, the only biologically pertinent values of the mutation rate, with the same results as those presented in the next subsection for continuously distributed populations [15, 22].

For work on stepping-stone models of finite length, the reader may consult [21] and references therein and the lists of the many papers of Maruyama in [12] and [16].

2.2. Continuous space, discrete time. We now pass formally from (1) to an integro (in space)-difference (in time) equation. In the stepping-stone model, deterministic variation of the population size in each colony is easy to arrange. For instance, for the simple uniform case (6), the random drift stage in each colony proceeds according to the classical Wright model of binomial sampling [15]. A rigorous derivation of the continuous space process, either as a limiting case of the stepping-stone model or directly is nontrivial; see [41, 42, 43].

Let us suppose the volume of an n-dimensional cube in the lattice is v, and define the migration and population densities through the relations $m(t, \mathbf{x}, \mathbf{y}) = v\tilde{m}(t, \mathbf{x}, \mathbf{y})$ and $N(t, \mathbf{x}) = v\rho(t, \mathbf{x})$. As $v \to 0$, we obtain from (1)

$$f(t+1, \mathbf{x}, \mathbf{y}) = (1-u)^2\Big\{\int\int m(t, \mathbf{z}, \mathbf{x}) m(t, \mathbf{w}, \mathbf{y}) f(t, \mathbf{z}, \mathbf{w}) d^n z\, d^n w + \int [2\rho(t, \mathbf{z})]^{-1} m(t, \mathbf{z}, \mathbf{x}) m(t, \mathbf{z}, \mathbf{y}) \times [1 - f(t, \mathbf{z}, \mathbf{z})] d^n z\Big\}, \qquad (11)$$

where we have dropped the tilde for notational simplicity. If the population density and migration are homogeneous and constant, we find from (6)

$$f(t+1, \mathbf{x}) = (1-u)^2\Big\{\int r(\mathbf{y}) f(t, \mathbf{x}-\mathbf{y}) d^n y + (2\rho)^{-1}[1 - f(t, \mathbf{0})] r(\mathbf{x})\Big\}, \qquad (12)$$

with

$$r(\mathbf{x}) = \int m(\mathbf{y}) m(\mathbf{x}+\mathbf{y}) d^n y. \qquad (13)$$

In (12) and (13), the spatial variables refer to separations, as in (4) and (6).

To separate the transient and equilibrium components of the probability of allelism, we assume we can write

$$f(t, \mathbf{x}) = F(t, \mathbf{x}) + f_\infty(\mathbf{x}), \qquad (14)$$

where $F(t, \mathbf{x}) \to 0$ as $t \to \infty$. From (12) and (14) we deduce at once

$$f_\infty(\mathbf{x}) = (1-u)^2\{\int r(\mathbf{y}) f_\infty(\mathbf{x}-\mathbf{y}) d^n y + s\, r(\mathbf{x})\}, \qquad (15)$$

$$F(t+1, \mathbf{x}) = (1-u)^2\Big\{\int r(\mathbf{y}) F(t, \mathbf{x}-\mathbf{y}) d^n y - (2\rho)^{-1} F(t, \mathbf{0}) r(\mathbf{x})\Big\}, \qquad (16)$$

where $s = [1 - f_\infty(\mathbf{0})]/(2\rho)$.

For a circular habitat, a detailed analysis of (15) and (16) and some biological discussion were presented in [20]. Other relevant material for finite habitats appears in [19] and [23].

To investigate the approach to equilibrium in an infinite habitat without boundaries, we first remove the factor $(1-u)^2$ in (16) by writing $F(t, \mathbf{x}) = -(1-u)^{2t}\varphi(t, \mathbf{x})$, and employ the Fourier transforms

$$\Phi(t, \mathbf{k}) = \int \varphi(t, \mathbf{x}) e^{-i\mathbf{k}\cdot\mathbf{x}} d^n x, \tag{17}$$

$$R(\mathbf{k}) = \int r(\mathbf{x}) e^{-i\mathbf{k}\cdot\mathbf{x}} d^n x. \tag{18}$$

Then (16) yields

$$\Phi(t+1, \mathbf{k}) = \left[\Phi(t, \mathbf{k}) - (2\rho)^{-1}\varphi(t, \mathbf{0})\right] R(\mathbf{k}). \tag{19}$$

Iterating (19) leads to

$$\Phi(t, \mathbf{k}) = \Phi(0, \mathbf{k})\left[R(\mathbf{k})\right]^t - (2\rho)^{-1}\sum_{n=0}^{t-1}\varphi(n, \mathbf{0})\left[R(\mathbf{k})\right]^{t-n}. \tag{20}$$

Therefore,

$$\begin{aligned}\varphi(t, \mathbf{x}) &= \int \Phi(t, \mathbf{k}) e^{i\mathbf{k}\cdot\mathbf{x}} \frac{d^n k}{(2\pi)^n} \\ &= I(t, \mathbf{x}) - (2\rho)^{-1}\sum_{n=0}^{t-1}\varphi(n, \mathbf{0}) G(t-n, \mathbf{x}),\end{aligned} \tag{21}$$

where

$$G(t, \mathbf{x}) = \int \left[R(\mathbf{k})\right]^t e^{i\mathbf{k}\cdot\mathbf{x}} \frac{d^n k}{(2\pi)^n}, \tag{22}$$

$$I(t, \mathbf{x}) = \int \varphi(0, \mathbf{y}) G(t, \mathbf{x}-\mathbf{y}) d^n y. \tag{23}$$

To solve for the time dependence in (21), we define the generating functions

$$\gamma(z,\mathbf{x}) = \sum_{t=1}^{\infty} \varphi(t,\mathbf{x})z^t, \tag{24}$$

$$\alpha(z,\mathbf{x}) = \sum_{t=1}^{\infty} G(t,\mathbf{x})z^t, \tag{25}$$

$$\beta(z,\mathbf{x}) = \sum_{t=1}^{\infty} I(t,\mathbf{x})z^t, \tag{26}$$

and obtain from (21)

$$\gamma(z,\mathbf{x}) = \beta(z,\mathbf{x}) - (2\rho)^{-1}\left[\varphi(0,\mathbf{0}) + \gamma(z,\mathbf{0})\right]\alpha(z,\mathbf{x}).$$

Solving this for $\gamma(z,\mathbf{0})$ and substituting, we find

$$\gamma(z,\mathbf{x}) = \beta(z,\mathbf{x}) - \frac{\alpha(z,\mathbf{x})\left[\varphi(0,\mathbf{0}) + \beta(z,\mathbf{0})\right]}{2\rho + \alpha(z,\mathbf{0})}. \tag{27}$$

If $r(\mathbf{x})$ is Gaussian and we rotate and scale coordinates so that its covariance matrix becomes the $n \times n$ unit matrix, then

$$G(t,\mathbf{x}) = (2\pi t)^{-n/2} e^{-|\mathbf{x}|^2/(2t)}. \tag{28}$$

The Central Limit Theorem implies that (28) holds as $t \to \infty$ for any $r(\mathbf{x})$ with finite covariance matrix. Tauberian Theorems [26, pp. 418–423] allow one to obtain the behavior of $\varphi(t,\mathbf{x})$ as $t \to \infty$ from that of $\gamma(z,\mathbf{x})$ as $z \to 1-$. In the special case of the stepping-stone model with $f(0,\mathbf{x}) = 0$, meaningful for identity by descent, the calculations have been carried out in one dimension by Malécot [21] and in both one and two dimensions for $u = 0$ and for $f(0,\mathbf{x}) = 0$ by Sawyer [46]. The complete solution for arbitrary initial condition appears in [40]. The unidimensional results in the next subsection agree with the corresponding ones in [40].

To illustrate the procedure, we note first that (22) and (25) yield

$$\alpha(z, \mathbf{x}) = z \int \frac{R(\mathbf{k}) e^{i\mathbf{k}\cdot\mathbf{x}}}{1 - zR(\mathbf{k})} \frac{d^n k}{(2\pi)^n}. \tag{29}$$

Let us assume $f(0, \mathbf{x}) = 0$ and $u = 0$. We shall see from (40) and (42) that $f_\infty(\mathbf{x}) = 1$ in one and two dimensions provided $u = 0$. Then (14) gives $\varphi(0, \mathbf{x}) = 1$, from which (22) and (23) allow us to infer $I(t, \mathbf{x}) = 1$. So (26) tells us that $\beta(z, \mathbf{x}) = z/(1 - z)$, and (27) simplifies to

$$\gamma(z, \mathbf{x}) = -1 + \frac{2\rho + \alpha(z, \mathbf{0}) - \alpha(z, \mathbf{x})}{(1 - z)[2\rho + \alpha(z, \mathbf{0})]}. \tag{30}$$

We now confine ourselves to one dimension, and consider first $x = 0$. From (25), (28), and a standard Tauberian Theorem [26, p. 423] we see that

$$\alpha(z, 0) \sim [2(1 - z)]^{-\frac{1}{2}} \quad \text{as } z \to 1-. \tag{30a}$$

Using (30a) in (30), we get

$$\gamma(z, 0) \sim 2\sqrt{2}\,\rho(1 - z)^{-\frac{1}{2}} \quad \text{as } z \to 1-.$$

Therefore, (24) and the same Tauberian Theorem inform us that [21]

$$\varphi(t, 0) \sim 2\rho\sqrt{\frac{2}{\pi t}},$$

in agreement with the continuous time result (67).

For $x \neq 0$, we calculate from (29)

$$\alpha(z, 0) - \alpha(z, x) = \frac{z}{\pi} \int_0^\infty \frac{(1 - \cos kx) R(k)}{1 - zR(k)} dk$$

$$\to \frac{2}{\pi} \int_0^\infty \frac{(1 - \cos kx) R(k) dk}{1 - R(k)}$$

as $z \to 1-$. We put $kx = \kappa$, let $x \to \infty$, and note that $R(k) = 1 - \frac{1}{2}k^2 + o(k^2)$ as $k \to 0$ to obtain

$$\alpha(z, 0) - \alpha(z, x) \sim \frac{4x}{\pi} \int_0^\infty \frac{\sin^2(\kappa/2)}{\kappa^2}\, d\kappa = x \qquad \text{(30b)}$$

for $z \to 1-$ and $x \to \infty$. Hence, (30), (30a), and (30b) yield for $x \gg 1$

$$\gamma(z, x) \sim (2\rho + x)\sqrt{2}\,(1 - z)^{-\frac{1}{2}} \quad \text{as } z \to 1-,$$

and the Tauberian Theorem leads to [21]

$$\varphi(t, x) \sim \sqrt{\frac{2}{\pi t}}\,(2\rho + x); \quad x \gg 1, \quad t \to \infty.$$

This also agrees with (67).

In two dimensions, some of Sawyer's results [46](see also [40]) can be deduced from (29), (30), and the same Tauberian Theorem. From (29), it is easy to see that

$$\alpha(z, \mathbf{0}) \sim -(2\pi)^{-1} \ln(1 - z) \quad \text{as } z \to 1-. \qquad \text{(30c)}$$

Inserting this into (30) gives

$$\gamma(z, \mathbf{0}) \sim -\frac{4\pi\rho}{(1 - z)\ln(1 - z)} \quad \text{as } z \to 1-,$$

whence [26, p. 423]

$$\varphi(t, \mathbf{0}) \sim 4\pi\rho/\ln t.$$

One can derive from (29) the formula

$$\alpha(z, \mathbf{0}) - \alpha(z, \mathbf{x}) \to \int \frac{(1 - \cos \mathbf{k} \cdot \mathbf{x})R(\mathbf{k})}{1 - R(\mathbf{k})} \frac{d^2k}{4\pi^2} \quad \text{as } z \to 1-$$

$$\sim \pi^{-1} \ln x \quad \text{as } x \equiv |\mathbf{x}| \to \infty. \qquad \text{(30d)}$$

Combining (30), (30c), and (30d) leads to the generating function

$$\gamma(z, \mathbf{x}) \sim -\frac{4\pi\rho + 2 \ln x}{(1 - z) \ln (1 - z)}$$

as $z \to 1-$ and $x \to \infty$. Applying again the Tauberian Theorem [26, p. 423], we find for $t \to \infty$, $x \to \infty$

$$\varphi(t, \mathbf{x}) \sim (4\pi\rho + 2 \ln x)/ \ln t.$$

Note that the ultimate rate of approach to complete homozygosis in two dimensions, $1/\ln t$, is much slower than the corresponding one-dimensional rate $t^{-\frac{1}{2}}$. In a practical evolutionary sense, the two-dimensional rate is so slow that total loss of genetic variability will not be very closely approximated.

In the presence of mutation ($u > 0$) or if $f(0, \mathbf{x}) \neq 0$, it is necessary to employ the general formula (27) instead of its special case (30). This more difficult analysis is presented in [40].

Let us turn now to the equilibrium problem (15). By analogy with (7), we use the Fourier transform [13]

$$\varphi_\infty(\mathbf{k}) = \int f_\infty(\mathbf{x}) e^{-i\mathbf{k}\cdot\mathbf{x}} d^n x$$

to solve (15). We find directly

$$f_\infty(\mathbf{x}) = s(1 - u)^2 \int \frac{d^n k}{(2\pi)^n} \frac{e^{i\mathbf{k}\cdot\mathbf{x}} R(\mathbf{k})}{1 - (1 - u)^2 R(\mathbf{k})}, \tag{31}$$

where $R(\mathbf{k})$ is given by (18). Since for all problems of biological interest $u \ll 1$ ($u \approx 10^{-5}$ at the genic level, $u \approx 10^{-9}$ at the nucleotide level), a general approximation to (31) can be developed. Owing to (5), $r(\mathbf{x})$ has only even moments. Rotating and scaling coördinates so that the covariance matrix of r (assumed to be finite for the present) becomes just the unit matrix, (18) yields

$$R(\mathbf{k}) = 1 - \frac{1}{2} k^2 + o(k^2) \text{ as } \mathbf{k} \to \mathbf{0}, \tag{32}$$

where $k^2 \equiv |\mathbf{k}|^2$. From (31) and (32), we expect as $u \to 0$ the main contribution to (31) to come from the neighborhood of $\mathbf{k} = \mathbf{0}$.

To explain the method, let us consider one dimension. Since $f_\infty(-x) = f_\infty(x)$, we may take $x \geqslant 0$. Then we have from (31) and (32)

$$f_\infty(x) = \pi^{-1}s(1-u)^2[J(x) + K(x)], \tag{33}$$

$$J(x) = \int_0^\infty \frac{\cos kx}{(1+\frac{1}{2}k^2) - (1-2u)}\, dk = \frac{\pi e^{-2\sqrt{u}\,x}}{2\sqrt{u}}, \tag{34}$$

$$K(x) = \int_0^\infty \left[\frac{1}{T(k) - (1-u)^2} - \frac{1}{(1+\frac{1}{2}k^2) - (1-2u)} \right] \cos kx dk, \tag{35}$$

$$T(k) = 1 + \frac{1}{2}k^2 + k^2 g(k), \tag{36}$$

where $T(k) = [R(k)]^{-1}$, $g(k) = o(1)$ as $k \to 0$, and (36) follows from (32). We write $K(x) = K_1(x) + K_2(x)$ by integrating from 0 to δ ($\delta > 0$) for K_1 and from δ to ∞ for K_2. For fixed $\delta > 0$, K_2 is finite for all $u \geqslant 0$. Hence, $K_2 = O(1)$ as $u \to 0$. Now, (35) and (36) give

$$K_1(x) = \int_0^\delta \frac{[u^2 - k^2 g(k)] \cos kx}{(\frac{1}{2}k^2 + 2u)[\frac{1}{2}k^2 + 2u + k^2 g(k) - u^2]}\, dk. \tag{37}$$

For any $\epsilon > 0$, no matter how small, there exists $\delta > 0$ such that $|g(k)| < \epsilon$ for $0 \leqslant k \leqslant \delta$. For sufficiently small u, (37) permits us to conclude

$$|K_1(x)| < 2\int_0^\delta \frac{u^2 + \epsilon k^2}{(\frac{1}{2}k^2 + 2u)^2}\, dk$$
$$< (\pi/4)(\sqrt{u} + 4\epsilon/\sqrt{u}) = o(1/\sqrt{u})$$

as $u \to 0$, where, to perform the integration, we replaced the upper limit by ∞. Therefore, we may express $f_\infty(x)$ in the form

$$f_\infty(x) = \frac{se^{-2\sqrt{u}\,x}}{2\sqrt{u}}\left[1 + h(u, x)\right], \tag{38}$$

where $h(u, x) = o(1)$ as $u \to 0$. Hence,

$$s = \frac{1}{2\rho}\left[1 - f_\infty(0)\right] = \frac{1}{2\rho}\left\{1 - \frac{s}{2\sqrt{u}}\left[1 + h(u, 0)\right]\right\}, \tag{39}$$

so

$$f_\infty(x) = \frac{e^{-2\sqrt{u}\,x}\left[1 + h(u, x)\right]}{1 + 4\rho\sqrt{u} + h(u, 0)}, \qquad x \geqslant 0. \tag{40}$$

As $u \to 0$, $f_\infty(x) \to 1$, i.e., if there is no mutation, the population ultimately becomes homozygous. Since the population is infinite, this complete loss of genetic variability is a consequence of geographical structure.

The result (40) was first given for Gaussian migration with the approximation $h = 0$ by Malécot [24]. If $g(k) = O(k^{1+\eta})$, $\eta > 0$, as $k \to 0$, as will be the case if the fourth moment of $r(x)$ is finite, for then $g(k) = O(k^2)$ as $k \to 0$, then $K_1(x) = O(1)$ as $u \to 0$, so $h(u, x) = O(\sqrt{u})$ as $u \to 0$, and we may, indeed, neglect $h(u, 0)$ in (40) for large ρ to obtain

$$f_\infty(x) \sim \frac{e^{-2\sqrt{u}\,x}}{1 + 4\rho\sqrt{u}}, \qquad x \geqslant 0. \tag{41}$$

If $h(u, x)$ does not approach zero as $u \to 0$ at least as fast as $\sqrt{u}$, then for very small u we may not be able to disregard $h(u, 0)$ in (40), though, of course, we can always drop the $h(u, x)$ in the numerator. In the simpler and biologically much more important situation $h = O(\sqrt{u})$, the formula [10, p. 83]

$$f_\infty(0) = \frac{1 + h(u,0)}{1 + 4\rho\sqrt{u} + h(u,0)}$$

$$= \frac{1}{\left[1 + 4\rho\sqrt{u} + h(u,0)\right]\left[1 - h(u,0)\right]} + O(u)$$

$$= \frac{1}{1 + 4\rho\sqrt{u}} + O(u) \quad \text{as} \quad u \to 0$$

for the homozygosity is worth noting.

In two dimensions, Malécot [13] has derived a formula for isotropic Gaussian migration and $x \equiv |\mathbf{x}| \gg 1$. For arbitrary migration with finite second moments, in terms of scaled and rotated coördinates we have [22]

$$f_\infty(\mathbf{x}) = \frac{K_0(2\sqrt{u}\,x) - K_0(\sqrt{2}\,x) + h(u,\mathbf{x})}{2\pi\rho - \frac{1}{2}\ln 2u + h(u,\mathbf{0})}, \tag{42}$$

in which K_0 is the modified Bessel function of the second kind of order zero and $h(u, x) = o(\ln u)$ as $u \to 0$. With the very weak assumption in (36) that $g(\mathbf{k}) = O(k^\eta)$, $\eta > 0$, as $k \to 0$, $h(u, \mathbf{x}) = O(1)$ as $u \to 0$, and h may be neglected in (42) for large ρ. Then the probability that a randomly selected individual in the population is homozygous is [13, 22]

$$f_\infty(\mathbf{0}) = \frac{1}{1 - 4\pi\rho/\ln 2u} + O\left[(\ln u)^{-2}\right] \tag{43}$$

as $u \to 0$. If $2\sqrt{u}\,x \gg 1$, (42) and (43) lead to [13, 22]

$$f_\infty(\mathbf{x}) \sim -\frac{f_\infty(\mathbf{0})}{\ln 2u}\sqrt{\frac{\pi}{\sqrt{u}\,x}}\; e^{-2\sqrt{u}\,x}, \tag{44}$$

a decay faster by $\sqrt{x}$ than the unidimensional one in (41). Just as in one dimension, we can verify easily from (42) that $f_\infty(\mathbf{x}) \to 1$ as $u \to 0$.

Since the integral in (31) is no longer singular as $u \to 0$, the three dimensional problem is more difficult. If $r(x)$ is spherically symmetric in the scaled coördinates, $r(x) = O(e^{-\alpha x})$, $\alpha > 0$, as $x \to \infty$, and

$$\int_0^{\infty} xr(x)\, dx < \infty,$$

one can show [25] that for $u \ll 1$, $f_\infty(x) \sim (\text{const.})\, e^{-2\sqrt{u}\,x}/x$ as $x \to \infty$. The decay rate should be compared to (44). Mutation is no longer required to maintain genetic diversity: $f_\infty(x) \nrightarrow 1$ as $u \to 0$. For less restrictive hypotheses, see [45].

Much more elaborate analysis is required if the second moments of $r(\mathbf{x})$ are not all finite. In one dimension, if $r(x) \sim b\gamma x^{-1-\gamma}$, $b > 0$ and $1 \leqslant \gamma < 2$, and some very weak and highly plausible technical assumptions are satisfied, it can be proved [25] that for $u \ll 1$ and $x \gg [a/(2u)]^{1/\gamma}$, $a = 2b\Gamma(1-\gamma)\cos(\gamma\pi/2)$,

$$f_\infty(x) \sim \frac{bs\gamma^2}{4u^2 x^{1+\gamma}}\,. \tag{45}$$

If $1 < \gamma < 2$, we may compute the heterozygosity [25]

$$1 - f_\infty(0) \sim 2\rho s \sim 2\rho\gamma a^{1/\gamma}(2u)^{(\gamma-1)/\gamma}\sin(\pi/\gamma). \tag{46}$$

2.3. Continuous space and time. To effect the passage from discrete to continuous time, we replace in (11) $t+1$ by $t + \Delta t$, u by $u\,\Delta t$, and multiply the second integral by Δt. This is done on the assumption that, with time measured in generations, the probabilities of mutation and reproduction are $u\,\Delta t$ and Δt. For clarity, we rewrite the migration function as $m(t, \mathbf{x}'; t + \Delta t, \mathbf{x})$ to exhibit its dependence on the time interval Δt. It will simplify the notation to use $2n$-dimensional vectors $\boldsymbol{\xi} = (\mathbf{x}, \mathbf{y})$, and to set $\mu(t, \boldsymbol{\xi}'; t + \Delta t, \boldsymbol{\xi}) = m(t, \mathbf{x}'; t + \Delta t, \mathbf{x})m(t, \mathbf{y}'; t + \Delta t, \mathbf{y})$. Now (11) reads

$$f(t + \Delta t, \boldsymbol{\xi}) = (1 - u\,\Delta t)^2 \int d^{2n}\xi'\, \mu(t, \boldsymbol{\xi}'; t + \Delta t, \boldsymbol{\xi}) \Big\{ f(t, \boldsymbol{\xi}') + (\Delta t)\big[2\rho(t, \mathbf{x}')\big]^{-1}\big[1 - f(t, \boldsymbol{\xi}')\big]\, \delta(\mathbf{x}' - \mathbf{y}')\Big\}, \tag{47}$$

where we have introduced the Dirac delta function in n dimensions. From (47) it follows that

$$\lim_{\Delta t\to 0} \frac{f(t+\Delta t, \xi) - f(t,\xi)}{\Delta t}$$

$$= \lim_{\Delta t\to 0} \int d^{2n}\xi' \mu(t, \xi'; t+\Delta t, \xi)) \left\{ \frac{f(t,\xi') - f(t,\xi)}{\Delta t} \right.$$

$$\left. -2uf(t,\xi') + \left[2\rho(t,\mathbf{x}')\right]^{-1}\left[1 - f(t,\xi')\right] \delta(\mathbf{x}' - \mathbf{y}')\right\}. \quad (48)$$

To deduce a partial differential equation from (48), we generalize the derivation of the Kolmogorov backward equation by Feller [26, pp. 321–323]. We know $0 \leqslant f(t,\xi) \leqslant 1$, suppose that $\rho(t,\mathbf{x})$ has a positive lower bound, and make Feller's diffusion assumptions. For all $\epsilon > 0$,

$$\lim_{\Delta t\to 0} \frac{1}{\Delta t} \int_{|\xi'-\xi|>\epsilon} \mu(t,\xi'; t+\Delta t, \xi)\, d^{2n}\xi' = 0, \quad (49a)$$

$$\lim_{\Delta t\to 0} \frac{1}{\Delta t} \int_{|\xi'-\xi|<\epsilon} (\xi_i' - \xi_i)\, \mu(t,\xi'; t+\Delta t, \xi)\, d^{2n}\xi' = M_i(t,\xi), \quad (49b)$$

$$\lim_{\Delta t\to 0} \frac{1}{\Delta t} \int_{|\xi'-\xi|<\epsilon} (\xi_i' - \xi_i)(\xi_j' - \xi_j) \times \mu(t,\xi'; t+\Delta t, \xi)\, d^{2n}\xi' = V_{ij}(t,\xi), \quad (49c)$$

with $i, j = 1, 2, \ldots, 2n$. If $f(t,\xi)$ has two continuous spatial derivatives, then (48) and (49) lead to [26, pp. 321–323]

$$\frac{\partial f}{\partial t} = -2uf + \sum_{i=1}^{2n} M_i(t,\xi) \frac{\partial f}{\partial \xi_i} + \frac{1}{2} \sum_{i,j=1}^{2n} V_{ij}(t,\xi) \frac{\partial^2 f}{\partial \xi_i \partial \xi_j}$$

$$+ \left[2\rho(t,\mathbf{x})\right]^{-1}\left[1 - f(t,\xi)\right] \delta(\mathbf{x} - \mathbf{y}). \quad (50)$$

It is argued in [16] and [22] that the normal derivatives of f should be set equal to zero on boundaries across which there is no migration.

For a special case of the above development, the reader may refer to [19]. Different methods were employed for various particular cases in [16, 22]. Since μ factors, if $M_i = 0$ for $i = 1, 2, \ldots, 2n$, then $V_{ij} = 0$ if $i \leqslant n$ and $j > n$ or $i > n$ and $j \leqslant n$. Then, with time- and space-independent second moments, scaling and rotation reduce (50) to [22]

$$\frac{\partial f}{\partial t} = -2uf + \frac{1}{2}\nabla_\xi^2 f + \left[2\rho(t, \mathbf{x})\right]^{-1}\left[1 - f(t, \xi)\right]\delta(\mathbf{x} - \mathbf{y}). \tag{51}$$

In (51), x_i and y_i are measured in terms of the corresponding standard deviations of m.

Perturbation theory was applied to (51) in [22] to prove that, for very high constant and uniform population density in a finite habitat of any shape and dimensionality, the rate of approach to equilibrium is $e^{-[2u+(2N)^{-1}]t}$, where N is the total population size. This is just the well-known Wright formula for a panmictic population. If $u = 0$, the population is asymptotically panmictic [22], and its genetic variability decays at the rate $e^{-t/(2N)}$. See [22] for additional discussion.

A detailed analysis of the covariance form of (51) for a line segment with constant and uniform population density was presented by Fleming and Su [16].

Let us confine ourselves to a homogeneous habitat in which migration depends only on the time interval and displacement: $m(t, \mathbf{x}'; t + \Delta t, \mathbf{x}) = m(\Delta t, \mathbf{x} - \mathbf{x}')$. Then, assuming $f(0, \mathbf{x}, \mathbf{y}) = f(0, \mathbf{x} - \mathbf{y})$, we may take $f(t, \mathbf{x}, \mathbf{y}) = f(t, \mathbf{z})$, where $\mathbf{z} = \mathbf{x} - \mathbf{y}$ is the n-dimensional vector of separation. Now (49b) implies $M_i = M_{i+n}$, $i \leqslant n$, and the drift terms cancel in pairs in (50). We have also, $V_{ii} = V_{i+n, i+n}$, $i \leqslant n$, and $V_{ij} = M_i M_j$, $i \leqslant n, j > n$ or $i > n, j \leqslant n$. Therefore, (50) reduces to

$$\frac{\partial f}{\partial t} = -2uf + \frac{1}{2}\sum_{i,j=1}^{n} C_{ij}\frac{\partial^2 f}{\partial z_i \partial z_j} + \left[\frac{1 - f(t, \mathbf{0})}{2\rho}\right]\delta(\mathbf{z}), \tag{52}$$

where $C_{ij} = 2(V_{ij} - M_i M_j)$, $i, j \leqslant n$, is twice the covariance matrix of m. From (13), a brief calculation shows that C_{ij} is the covariance matrix of r. Diagonalizing C_{ij} and measuring z_i in terms of the corresponding standard deviation of r gives

$$\frac{\partial f}{\partial t} = -2uf + \frac{1}{2}\nabla_z^2 f + \left[\frac{1 - f(t, \mathbf{0})}{2\rho}\right]\delta(\mathbf{z}). \tag{53}$$

The divergence theorem permits us to see directly that (53) has no solution in two or more dimensions [16, 22] due to the violent singularity at the origin. The complete solution of (53) on a circle appears in [20].

The solution of (53) on the infinite line is quite instructive [22]. With the decomposition (14) we obtain from (53)

$$-2uf_\infty + \frac{\sigma^2}{2}\frac{d^2 f_\infty}{dx^2} + \left[\frac{1 - f_\infty(0)}{2\rho}\right]\delta(x) = 0, \tag{54}$$

$$\frac{\partial F}{\partial t} = -2uF + \frac{\sigma^2}{2}\frac{\partial^2 F}{\partial x^2} - \frac{F(t, 0)}{2\rho}\delta(x) = 0. \tag{55}$$

We have reintroduced the variance σ^2 of r in order to express the separation x in arbitrary units. If we replace x by x/σ and ρ by $\rho\sigma$ in (41), we obtain the solution of (54). Thus, the approximation for $u \ll 1$ for discrete time is the exact solution of the continuous time problem. The transient component of the probability of nonallelism is

$$H(t, x) = -F(t, x) = e^{-2ut}\varphi(t, x). \tag{56}$$

Substituting (56) into (55) yields

$$\frac{\partial \varphi}{\partial t} = \frac{\sigma^2}{2}\frac{\partial^2 \varphi}{\partial x^2} - \frac{\varphi(t, 0)}{2\rho}\delta(x). \tag{57}$$

To remove dimensions from (57), let $x = a\xi$, $t = T\tau$, and $\psi(\tau, \xi) = \varphi(t, x)$. From (57) we infer that the natural units of length and time are $a = 2\rho\sigma^2$ and $T = (2\rho\sigma)^2$. Then

$$\frac{\partial\psi}{\partial\tau} = \frac{1}{2}\frac{\partial^2\psi}{\partial\xi^2} - \psi(\tau, 0)\,\delta(\xi), \tag{58}$$

with the solution [22]

$$\psi(\tau, \xi) = \int_0^\infty \psi(0, \eta)\, g(\tau; \xi, \eta)\, d\eta, \quad \xi \geqslant 0, \tag{59}$$

$$g(\tau; \xi, \eta) = (2\pi\tau)^{-\frac{1}{2}}\left[e^{-(\xi+\eta)^2/(2\tau)} + e^{-(\xi-\eta)^2/(2\tau)}\right]$$
$$- e^{\frac{1}{2}\tau+\xi+\eta}\left\{1 - \operatorname{erf}\left[\sqrt{\tau/2} + (\xi + \eta)/\sqrt{2\tau}\,\right]\right\}. \tag{60}$$

The error function is

$$\operatorname{erf} x = \frac{2}{\sqrt{\pi}}\int_0^x e^{-y^2}\, dy, \tag{61}$$

and $\psi(\tau, -\xi) = \psi(\tau, \xi)$. Note that (14) and (56) give the initial value

$$\psi(0, \xi) = H(0, x) = f_\infty(x) - f(0, x). \tag{62}$$

The most important aspect of the solution is its behavior as $\tau \to \infty$. From (59) and (60) [22],

$$\psi(\tau, \xi) \sim \sqrt{2/(\pi\tau^3)}\,(1 + \xi)\int_0^\infty \psi(0, \eta)(1 + \eta)\, d\eta, \tag{63}$$

provided the integral in (63) converges. This approximation holds if $\sqrt{\tau} \gg 1, \xi, l$, where l is the largest (dimensionless) length in $\psi(0, \xi)$. If, for the moment, we consider identity by descent, we may take $f(0, x) = 0$ for an initially noninbred population. Hence, (41) and (62) imply

$$\psi(0, \xi) = \frac{e^{-2\sqrt{uT}\,\xi}}{1 + 2\sqrt{uT}}, \tag{64}$$

and the result of integrating (63) reads

$$\psi(\tau, \xi) \sim \sqrt{\frac{2}{\pi\tau^3}}\ \frac{1+\xi}{4uT}; \quad u > 0,\ \sqrt{\tau} \gg 1, \xi, 1/\sqrt{uT}\,. \tag{65}$$

If $u = 0$, and we still suppose $f(0, x) = 0$, then (64) informs us that $\psi(0, \xi) = 1$. The divergence of the integral in (63) causes the u^{-1} behavior in (65). However, using (60) and (61), integration by parts allows us to evaluate (59):

$$\psi(\tau, \xi) = \operatorname{erf}(\xi/\sqrt{2\tau}\,) + e^{\frac{1}{2}\tau+\xi}\left[1 - \operatorname{erf}\left(\sqrt{\tau/2} + \xi/\sqrt{2\tau}\,\right)\right]. \tag{66}$$

For $\sqrt{\tau} \gg 1, \xi$, we find from (66)

$$\psi(\tau, \xi) \sim \sqrt{\frac{2}{\pi\tau}}\ (1 + \xi), \tag{67}$$

which decays only at the rate $\tau^{-\frac{1}{2}}$ instead of the much faster rate $\tau^{-3/2}$ in (63) and (65). See [21] for a different derivation of (67). The significance of these results is enhanced by the observation, based on (41), that if $u = 0$ then $f_\infty(x) = 1$, i.e., the population ultimately becomes completely homozygous, and $H(t, x)$ is asymptotically a measure of its genetic variability.

3. MIGRATION AND SELECTION

We shall now consider the effect of spatial variation in natural selection. The type of model discussed in this section has been employed to estimate the intensity of natural selection [27], and pertains directly to speciation. We suppose that the population density is sufficiently high to allow us to neglect random drift, and thus deal only with the deterministic variation of gene frequencies in space and time. Refer to [38, 39, 57] for the effect of random drift.

Due to the essential nonlinearity of natural selection, a discrete-space, discrete-time model analogous to the one in Section

2.1 appears to have only numerical utility. A continuous-space, discrete-time model like that in Section 2.2 serves to illuminate the assumptions, and permits an intuitive passage to the fully continuous problem (cf. Section 2.3), to which almost all analysis has been devoted, as an approximation [27, 28]. Fisher [29, 30] wrote down the diffusion equation for very simple situations directly. We shall rely on the detailed general diffusion formulation in [31], but consider only its special case with isotropic, homogeneous, and time- and genotype-independent migration, and constant and uniform population density. As in [31], we restrict ourselves to a single diallelic locus in the absence of mutation, and posit that the monoecious population is locally in Hardy-Weinberg proportions. The removal of these assumptions is straightforward and biologically worthwhile, but greatly increases the difficulty of the mathematics. See [54, 55, 56]. Finally, we suppose that the degree of dominance (essentially, the ratio of the differences between the selection coefficients of the heterozygote and the two homozygotes) is independent of location.

Let the genotypes AA, Aa, aa have Malthusian parameters $sg(\mathbf{x})$, $hsg(\mathbf{x})$, and $-sg(\mathbf{x})$, respectively. If σ^2 is the variance per generation of migration, then the frequency $p(\mathbf{x}, t)$ of A satisfies [31]

$$\frac{\partial p}{\partial t} = \frac{\sigma^2}{2} \nabla^2 p + sg(\mathbf{x})f(p), \tag{68a}$$

$$f(p) = p(1-p)(1+h-2hp), \tag{68b}$$

$$\frac{\partial p}{\partial n} = 0 \quad \text{on the boundaries.} \tag{68c}$$

The notation in (68c) signifies the normal derivative, and (68c) is the appropriate condition if there is no migration across the boundaries.

We are concerned with the existence and characterization of spatially-varying equilibria, called clines, in heterogeneous environments. To model this environmental diversity, the direction of natural selection, and hence the sign of $g(\mathbf{x})$, will be assumed to reverse at least once in the habitat. The parameter s specifies the intensity of natural selection, and h determines the degree of dominance. The critical points of (68) are the zeroes of $f(p)$, viz., $p = 0, 1$, and $p^* \equiv (1 + h)/(2h)$. Since our interest is confined to solutions satisfying $0 \leqslant p(\mathbf{x}, t) \leqslant 1$, excluding overdominance ($|h| > 1$) will simplify matters by removing the critical point p^* from the interval $(0, 1)$. The phenomena in $[0, p^*]$ and $[p^*, 1]$ with $|h| > 1$ are qualitatively the same as the ones we shall explore in $[0, 1]$ for $|h| \leqslant 1$ [31].

Dirichlet boundary conditions are appropriate wherever the population under consideration is in contact with a much larger one. For instance, a linear habitat with the allelic frequency fixed at p_b could model a peninsula or archipelago projecting from a continent. If $p_b = 0$ or 1, this is equivalent to a Neumann problem (68) with one of the alleles lethal outside the habitat. If $0 < p_b < 1$, however, since $p(\mathbf{x}, t) = p_b$ is not a solution of (68a), it is clear that at least one spatially-varying solution must exist. Hence, the possible questions here are of less biological interest than for (68).

3.1. Linear habitat without boundaries. The one-dimensional case applies to two quite distinct biological circumstances. Organisms are frequently restricted to the immediate neighborhood of a river, seashore, etc., or may live in a river or narrow valley. Alternatively, the distribution of the population may really be two-dimensional, but only one of the coördinates may have genetic relevance. This would happen, e.g., if only latitude matters on a geographical scale, or locally if individuals exist at various elevations on a mountain range.

If the habitat may be treated as infinitely long, we seek solutions of

$$\frac{\partial p}{\partial t} = \frac{\sigma^2}{2}\frac{\partial^2 p}{\partial x^2} + sg(x)f(p), \tag{69a}$$

$$\frac{\partial p}{\partial x} \to 0 \quad \text{as} \quad |x| \to \infty. \tag{69b}$$

Of course, the trivial solutions $p(x, t) = 0, 1$ always exist. For the special cases of no dominance ($h = 0$) and A recessive ($h = -1$), with $g(x)$ equal to a negative constant for $x < 0$, and a positive one for $x > 0$, Haldane [27] derived explicitly the unique nontrivial equilibrium solutions of (69). More generally, for $h = 0$ and $g(x)$ subject only to the conditions $\operatorname{sgn} g(x) = \operatorname{sgn} x$ for sufficiently large $|x|$, and that $g(x)$ be *not* integrable as $|x| \to \infty$, Conley [32] has shown that a nontrivial equilibrium always exists. His proof holds for arbitrary dominance, and the theorem covers the existence of the clines studied by Haldane [27] and Fisher [30]. Consult [53] for more on existence, uniqueness, and stability.

We now turn to the interesting particular case of equally intense selection in opposite directions in the two halves of the habitat, $g(-x) = -g(x)$, $g(x) > 0$ for $x > 0$, and no dominance, $h = 0$. We have $f(p) = p(1-p) = f(1-p)$.

Let us show first that at least one symmetric equilibrium solution, i.e., one satisfying $\hat{p}(-x) = 1 - \hat{p}(x)$, exists. Such a cline is sketched in Figure 1. It is trivial to verify that if $p(x, t)$ is a solution of (69), so is $\tilde{p}(x, t) \equiv 1 - p(-x, t)$. If $p(x, 0) = 1 - p(-x, 0)$, then $\tilde{p}(x, 0) = p(x, 0)$, and therefore by uniqueness for the initial value problem, $p(x, t) = \tilde{p}(x, t) = 1 - p(-x, t)$. Hence, there exists at least one symmetric cline $\hat{p}(x)$.

Next, we prove that there is exactly one symmetric equilibrium. For, if there are two, $p_1(x)$ and $p_2(x)$, then (69) shows that

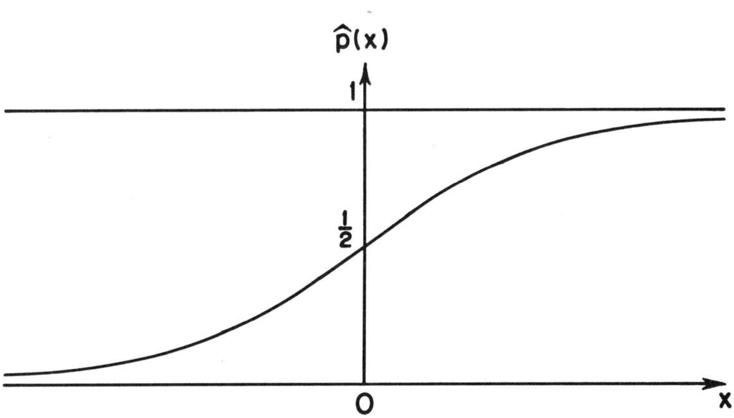

FIG. 1. Typical symmetric cline.

$P(x) \equiv p_1(x) - p_2(x)$ satisfies

$$\frac{\sigma^2}{2}\frac{d^2P}{dx^2} = sg(x)[p_1(x) + p_2(x) - 1]P(x). \tag{70}$$

Consider (70) on $x \geqslant 0$. We know that $P(0) = P(\infty) = 0$, and (70) informs us that $\operatorname{sgn} P''(x) = \operatorname{sgn} P(x)$. Therefore, $P(x) \equiv 0$, as required for uniqueness.

To investigate the stability of the symmetric solution $\hat{p}(x)$, we put $p(x, t) = \hat{p}(x) + \epsilon(x, t)$, and deduce from (69) the linearized equation

$$\frac{\partial \epsilon}{\partial t} = \frac{\sigma^2}{2}\frac{\partial^2 \epsilon}{\partial x^2} + sg(x)[1 - 2\hat{p}(x)]\epsilon \tag{71}$$

for $\epsilon(x, t)$. Separating variables, we see that $\epsilon(x, t) = \psi(x)e^{-\lambda t}$, with

$$\mathcal{H}\psi = \lambda\psi, \quad -\infty < x < \infty \tag{72a}$$

$$\mathcal{H} = -\frac{\sigma^2}{2}\frac{d^2}{dx^2} + U(x), \tag{72b}$$

$$U(x) = sg(x)[2\hat{p}(x) - 1]. \tag{72c}$$

Since $\epsilon(x, t)$ must be bounded, (72) is just a standard unidimensional quantum mechanics problem with energy λ and potential $U(x)$, sketched in Figure 2 for $g(\infty) < \infty$. Stability ($\lambda \geqslant 0$) is now obvious since $U(x) \geqslant 0$ implies $\mathcal{H}$ is a positive operator. If $g(\infty) = \infty$, the spectrum is discrete, and the asymptotic approach to equilibrium is given by $e^{-\lambda_0 t}$, where λ_0 is the smallest eigenvalue. That this is still so with $g(\infty) < \infty$ may be seen as follows. Inscribe the square well $U_S(x) \geqslant U(x)$ as shown in Figure 2. It is well known that every square well with equally high sides has at least one bound state [33, pp. 37–42]. Call the lowest of these $\psi_S(x)$, with eigenvalue $\lambda_S < sg(\infty)$. With the customary inner product notation to indicate integration, we find, with

$$\mathcal{H}_S = -\frac{\sigma^2}{2}\frac{d^2}{dx^2} + U_S(x), \tag{73}$$

$$\lambda_0 = (\psi_0, \mathcal{H}\psi_0) \leqslant (\psi_S, \mathcal{H}\psi_S) \leqslant (\psi_S, \mathcal{H}_S\psi_S) = \lambda_S. \tag{74}$$

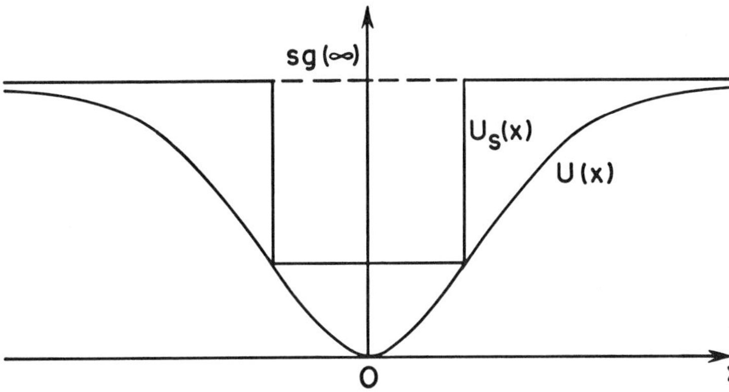

FIG. 2. Potential and square well for equation (72).

Thus, there is a nonempty discrete spectrum in $0 < \lambda < sg(\infty)$ and a continuous spectrum for $\lambda \geqslant sg(\infty)$.

We can now return to (69) to study the order of magnitude of the width of the cline $\hat{p}(x)$, defined by $w \approx 1/p'(0)$, and the rate of approach to equilibrium. The scaling argument for w applies for any h, but the results for the rate λ_0 have been justified above only for $h = 0$ and symmetric $\hat{p}(x)$, and may fail for $h \neq 0$. Suppose the environmental length scale is d and

$$g(x) = \begin{cases} -1, & x < -d/2, \\ -(-2x/d)^{\mu}, & -d/2 < x < 0, \\ +(2x/d)^{\mu}, & 0 < x < d/2, \\ +1, & x > d/2, \end{cases}$$

where $\mu \geqslant 0$. We set $x = w\xi$ and $t = T\tau$, and determine the length and time scales w and T so as to reduce (69) to dimensionless form for $|x| < d/2$. With $l = \sigma/\sqrt{s}$ [28] and $w = 2^{-\mu/(\mu+2)}w_0$, we find $\bar{p}(\xi, \tau) \equiv p(x, t)$ satisfies [31]

$$\frac{\partial \bar{p}}{\partial \tau} = \frac{1}{2}\frac{\partial^2 \bar{p}}{\partial \xi^2} + \xi^{\mu} f(\bar{p}), \quad 0 < \xi < d/(2w), \tag{75}$$

$$w_0 = (l^2 d^{\mu})^{1/(\mu+2)} = l(d/l)^{\mu/(\mu+2)} = d(l/d)^{2/(\mu+2)}, \tag{76}$$

and $T = w^2/\sigma^2$.

The factor $2^{-\mu/(\mu+2)}$ decreases from 1 to $\frac{1}{2}$ as μ increases from 0 to ∞, and hence has no bearing on the scale of length. The most rapid variation of $\hat{p}(x)$ clearly occurs for the step environment, $\mu = 0$, for which $w_0 = l$. So, for arbitrary μ, $w_0 \geqslant l$, and we must restrict (76) to $d \geqslant l$, which implies $l \leqslant w_0 \leqslant d$. This is consistent because (75) then encompasses $0 < \xi < \frac{1}{2}$. With $d < l$, (75) does not include all the region of significant variation of gene frequency, but we still have $w_0 \approx l$ [28]. If $\mu = \infty$, the regions where a particular allele is advantageous and where it is deleterious are separated by a neutral belt of width d, leading to the longest scale $w_0 = d$ for the cline $\hat{p}(x)$. The formula $w_0 = (l^2 d)^{1/3}$ for $\mu = 1$ and the time scales for $\mu = 0, 1$ were given recently by May, Endler, and McMurtrie [34]. These authors also compared their widths and Slatkin's [28] to the available experimental data. The gentler the environmental variation (increasing μ), the wider will be the cline and the slower the approach to equilibrium. If $d < l$, $T = 1/s$, the shortest characteristic time, while for $d \geqslant l$ and $\mu = 1, \infty$, $T = [d/(2s\sigma)]^{2/3} = [d/(2l)]^{2/3}/s$, $T = [d/(2\sigma)]^2$. The relaxation time for the step environment depends only on selection in a dimensionally obvious way because the only length, σ, clearly cannot enter. It is intuitively reasonable that the characteristic time within the neutral belt for $\mu = \infty$ should be independent of the selection intensity.

3.2. At least one allele favored in a finite region. Frequently at least one of the alleles will be favored in a region too small to be treated as infinite. We suppose the allele A is advantageous ($g > 0$) in a region of linear dimension b, and define the dimensionless coordinate $\boldsymbol{\xi} = \mathbf{x}/b$. Redefining g and ∇^2 to refer to ξ and dropping the caret from the cline $\hat{p}(\boldsymbol{\xi})$, we infer from (68) that $p(\boldsymbol{\xi})$ is the solution of

$$\nabla^2 p + k^2 g(\xi) f(p) = 0 \tag{77}$$

with Neumann boundary conditions. The crucial dimensionless parameter for the problem is $k^2 = 2sb^2/\sigma^2 = 2b^2/l^2$.

We consider first an infinite habitat, and rely on theorems established by Conley [32] for $h = 0$ which generalize to $-1 < h$

$\leqslant 1$. In one dimension, we posit that $g(\xi)$ is negative and nonintegrable as $\xi \to \pm\infty$, and $g(\xi) > 0$ in some interval. Then for sufficiently large k a cline exists which tends to zero as $|\xi| \to \infty$. By symmetry, this theorem also covers the corresponding problem on the half-line $\xi > 0$: $g(\xi) > 0$ in $[0, 1)$ and $g(\xi)$ negative and nonintegrable as $\xi \to \infty$. A typical cline with $g(\xi) > 0$ in $[0, 1)$ and $g(\xi) < 0$ for $\xi > 1$ is sketched in Figure 3. Reflecting it about $\xi = 0$ produces the environmental pocket discussed by Conley [32]. In two dimensions, we suppose [32] selection is isotropic, $g(\boldsymbol{\xi}) = g(\xi)$, $\xi = |\boldsymbol{\xi}|$, $g(\xi) > 0$ for $\xi < 1$, and $\xi g(\xi)$ is negative and nonintegrable as $\xi \to \infty$. Again, for sufficiently large k a cline exists which tends to zero as $\xi \to \infty$. If $g(\xi) < 0$ for $\xi > 1$, the qualitative picture remains that in Figure 3.

Uniqueness and stability have yet to be investigated in general. Numerical calculations [28, 35] lead us to expect that at least one cline will be stable.

Let us assume now that the habitat is finite, the situation studied by Fleming [36]. We may confine ourselves to the case

$$\int g(\boldsymbol{\xi})d^n\xi \leqslant 0, \tag{78}$$

for otherwise we substitute $q(\boldsymbol{\xi}) = 1 - p(\boldsymbol{\xi})$ in (77). We still suppose $-1 < h \leqslant 1$. If equality holds in (78), a cline always exists [36]. If the inequality is strict, a cline exists whenever k^2

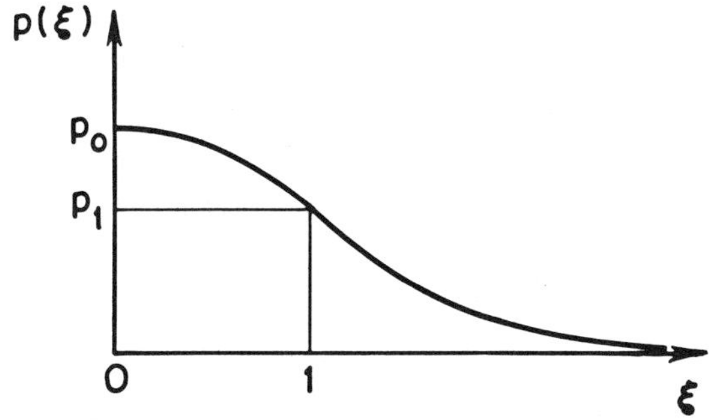

FIG. 3. Typical cline in a semi-infinite habitat.

exceeds κ^2, the smallest positive eigenvalue of (77) linearized around $p = 0$ [36]. For the one-dimensional symmetric case, $g(-\xi) = -g(\xi)$, $p(-\xi) = 1 - p(\xi)$, there exists exactly one cline, and it is stable [36]. Fleming has also shown that with equality in (78) the trivial solutions $p = 0, 1$ are both unstable, while with inequality $p = 1$ is always unstable, but $p = 0$ is stable for $k < \kappa$ and unstable for $k > \kappa$; in one dimension, instability of both trivial solutions implies stability of the cline [36].

The general conclusion from the above results is that a cline is maintained if, due to some combination of intense natural selection, weak migration, and high environmental heterogeneity, k exceeds a critical value κ. Usually, but not always, κ is of order unity. A few examples are instructive.

We take $g(\xi) = 1$, $0 \leqslant \xi < 1$; $g(\xi) = -\alpha^2$, $\xi > 1$. For $-1 < h \leqslant 1$, a sufficient condition for the existence of a cline in the semi-infinite habitat is $k_0 > \tan^{-1}\alpha$, where $k_0 = (1 + h)^{\frac{1}{2}}k$ [31]. If $-\frac{1}{3} \leqslant h \leqslant 1$, the condition is also necessary, and the cline is unique [31]. The critical value $\kappa = \tan^{-1}\alpha$ is of order unity unless $\alpha \ll 1$, in which case $\kappa \approx \alpha$. If $\alpha \gg 1$, $\kappa \approx \pi/2$, the smallest value of k_0 required for maintaining A in the population if it is lethal for $\xi > 1$. The greatest gene frequency $p_0(k) = p(0)$ in the cline is plotted for $\alpha = 1$ and $h = 0$ in Figure 4. Note the rapid rise of the curve from zero to one.

If A is recessive, $h = -1$, the general theory in [32, 36] does not apply. For the step environment of the previous paragraph, it was shown in [31] that for $k < \kappa$ there are only the trivial solutions, while for $k > \kappa$ there are an even (nonzero) number of clines. No explicit formula exists for κ. Numerical calculations indicate that the number of clines for $k > \kappa$ is two [31]. The maximum gene frequency $p_0(k)$, which completely characterizes the cline, is plotted for $\alpha = 1$ in Figure 5. A simple model in [31] and the intuitive argument that we should have $p_0(k) \to 1$ as $k \to \infty$ lead us to expect that the stable cline corresponds to the upper branch in Figure 5. If so, then $p_0(k)$ is greater than about $\frac{1}{2}$, and stable small clines do not exist. Intuitively, this happens because now $f(p) = 2p^2(1 - p)$, hence, for sufficiently small p and fixed k, quadratic selection is too weak to balance linear migration.

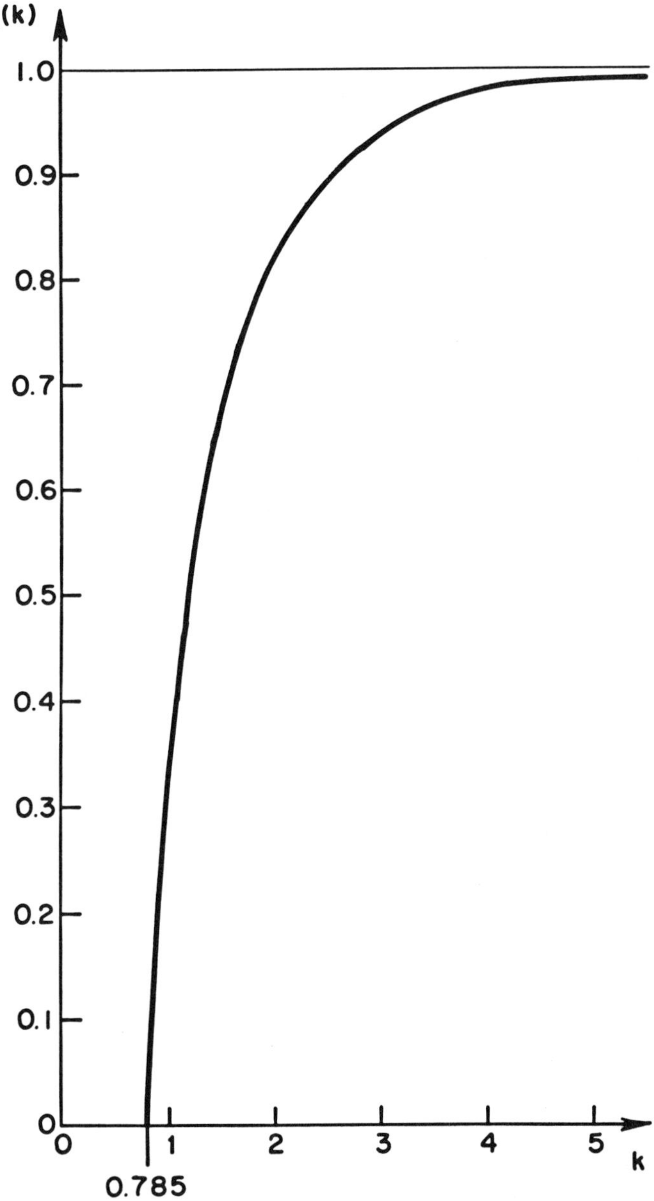

FIG. 4. The maximum gene frequency in the semi-infinite cline without dominance for $\alpha = 1$.

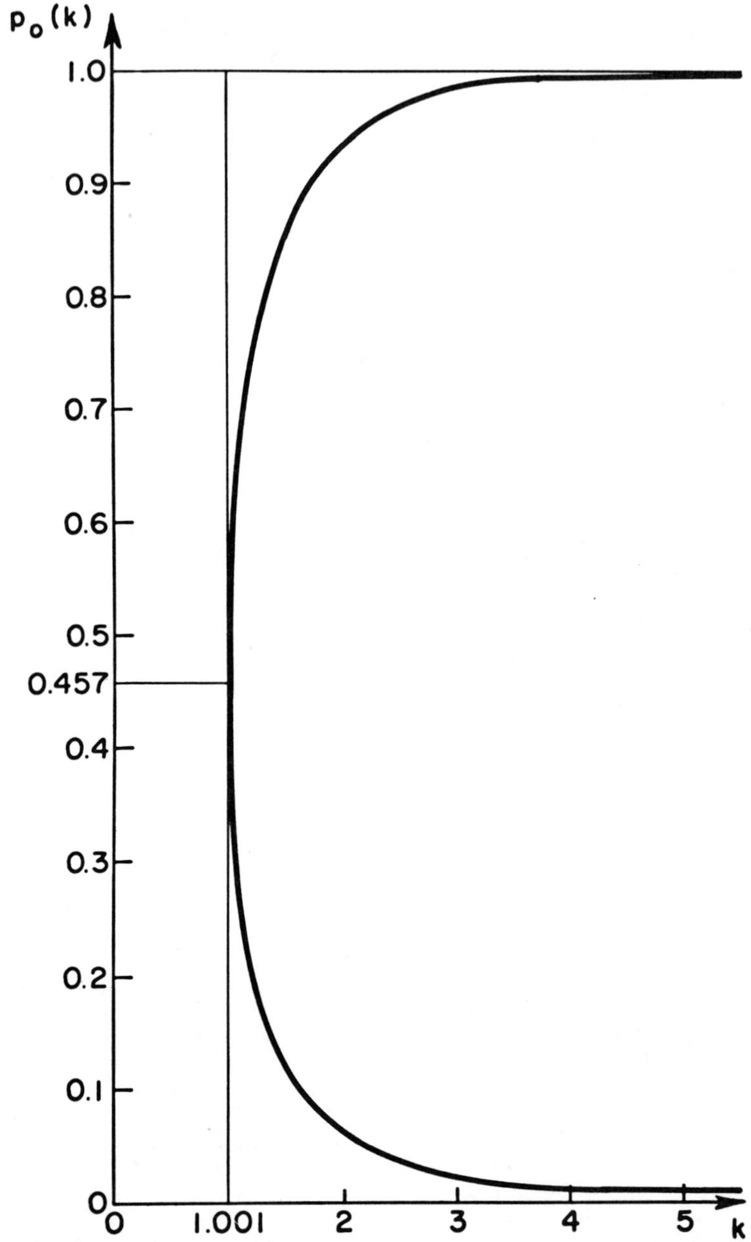

FIG. 5. The maximum gene frequency in the semi-infinite cline with A recessive for $\alpha = 1$.

As an example of a finite habitat, consider the line segment $[-1, \gamma]$ with $g(\xi) = 1$ for $-1 \leqslant \xi < 0$ and $g(\xi) = -\alpha^2$ for $0 < \xi \leqslant \gamma$ [31]. The integral in (78) is $1 - \alpha^2\gamma$, so, if $\alpha^2\gamma = 1$, a nontrivial solution always exists. If $\alpha^2\gamma > 1$, a sufficient condition for the existence of a cline is that $k > \kappa$, where κ is the unique root of $\kappa = \tan^{-1}[\alpha \tanh(\alpha\gamma\kappa)]$ [31]. Observe that $\kappa \leqslant \tan^{-1}\alpha$. As expected, if A is deleterious only in a finite region, the critical value is lower than for the semi-infinite cline.

For other examples in one and two dimensions and discussion, the reader may consult [31]. The main qualitative conclusion from the analyses is that the cline mechanism does not permit a population to carry many alleles at low frequencies. Thus, whatever flexible response the population has to environmental change in time exists primarily by virtue of alleles at high frequencies. The action of selection in a geographically heterogeneous environment in maintaining genetic variability is quite different from that of mutation, either in the mutation model of Section 2 or when mutation opposes purifying selection [7, pp. 258–262, 264–267].

3.3. Some extensions. A two-locus cline model has been recently studied by Slatkin [37]. The main results are that if linkage is tight (recombination frequency of the same order as the selection intensity or smaller), it has an appreciable effect on gene frequencies, and a substantial amount of linkage disequilibrium is present in the cline. Since population densities are frequently low, random drift may have significant effects in clines even if the total population size is large. Refer to [38, 39] for numerical work and to [57] for a diffusion model and its dimensional analysis. Especially interesting open problems remaining concern clines with multiple alleles.

Note added in proof: Since this chapter was completed, much work has been done on the problems discussed here. The theory of migration and selection with multiple niches, described and reviewed in [44, Ch. 6], provides useful background.

Sawyer has the most detailed and precise results on the equilibrium behavior of the stepping-stone model [45]. He also elucidates the underlying probabilistic structure of the model and studies the convergence to equilibrium [46]. The approach to equilibrium in finite populations is examined in [47]. Sawyer has derived interesting results concerning the spread of alleles [48], and has also investigated stepping-stone models in a "tree" such as a river system [49]. For a birth-and-death version of the stepping-stone model, consult new work of Malécot [50, 51]. Gregorius has formulated and studied migration, mutation, and random drift in a continuously distributed plant population [52].

Fife and Peletier [53] establish existence, uniqueness, and global stability for a cline in an infinite linear habitat with different alleles favored far away to the right and left. The effects of a geographical barrier and of a discontinuity in the migration rate are investigated in [54]; those of asymmetric migration and of a discontinuity in the population density in [55]. Consult [56] for a detailed mathematical exploration of the general case. A diffusion model and some results for random genetic drift in a cline appear in [57].

We did not discuss in this paper the interesting and important problem of the spread of a favored allele, described by (68) with $g(\mathbf{x}) = 1$. It was treated first by Fisher [58] and Kolmogorov, Petrovski, and Piskunov [59]. See [60–65] for recent work and references.

Acknowledgement: It is a pleasure to thank Professors Charles Conley, Bryce McLeod, Stanley Sawyer, and Stephen Wainger for helpful communication.

REFERENCES

1. T. Nagylaki, "Polymorphisms in cyclically-varying environments," *Heredity*, **35** (1975), 67–74.
2. S. Wright, *Evolution and the Genetics of Populations*, vol. 2, The University of Chicago Press, Chicago, 1969.
3. S. Karlin and J. L. McGregor, "Application of method of small parameters to multi-niche population genetic models," *Theor. Pop. Biol.*, **3** (1972), 186–209.
4. S. Sawyer, "Branching diffusion processes in population genetics," *Advances in Appl. Prob.*, **8** (1976), 659–689.

5. *Webster's Seventh New Collegiate Dictionary*, G. and C. Merriam Company, Springfield, Mass., 1972.
6. J. F. Crow, *Genetics Notes*, 6th ed., Burgess Publishing Company, Minneapolis, 1966.
7. J. F. Crow and M. Kimura, *An Introduction to Population Genetics Theory*, Harper and Row, New York, 1970.
8. W. J. Ewens, *Population Genetics*, Methuen, London, 1969.
9. M. Kimura and T. Ohta, *Theoretical Aspects of Population Genetics*, Princeton University Press, Princeton, 1971.
10. G. Malécot, *The Mathematics of Heredity*, Freeman, San Francisco, 1969.
11. P. A. P. Moran, *The Statistical Processes of Evolutionary Theory*, Clarendon Press, Oxford, 1962.
12. J. Felsenstein and B. Taylor, *A Bibliography of Theoretical Population Genetics*, National Technical Information Service, U.S. Department of Commerce, Springfield, VA, 1974.
13. G. Malécot, "Identical loci and relationship," *Proc. Fifth Berkeley Symp. Math. Stat. Prob.*, **4** (1967), 317–332.
14. M. Kimura and G. H. Weiss, "The stepping stone model of population structure and the decrease of genetic correlation with distance," *Genetics*, **49** (1964), 561–576.
15. G. H. Weiss and M. Kimura, "A mathematical analysis of the stepping stone model of genetic correlations," *J. Appl. Prob.*, **2** (1965), 129–149.
16. W. Fleming and C.-H. Su, "One dimensional migration models in population genetics theory," *Theor. Pop. Biol.*, **5** (1974), 431–449.
17. G. Malécot, "Quelques schémas probabilistes sur la variabilité des populations naturelles," *Ann. Univ. Lyon, Sciences Sec. A*, **13** (1950), 37–60.
18. M. Kimura, " 'Stepping Stone' model of population," *Ann. Rept. Nat. Inst. Genet., Japan*, **3** (1953), 62–63.
19. T. Maruyama, "The rate of decrease of heterozygosity in a population occupying a circular or a linear habitat," *Genetics*, **67** (1971), 437–454.
20. T. Nagylaki, "Genetic structure of a population occupying a circular habitat," *Genetics*, **78** (1974), 777–790.
21. G. Malécot, "Heterozygosity and relationship in regularity subdivided populations," *Theor. Pop. Biol.*, **8** (1975), 212–241.
22. T. Nagylaki, "The decay of genetic variability in geographically structured populations," *Proc. Nat. Acad. Sci. USA*, **71** (1974), 2932–2936.
23. T. Maruyama, "The rate of decrease of genetic variability in a two-dimensional continuous population of finite size," *Genetics*, **70** (1972), 639–651.
24. G. Malécot, "Decrease of relationship with distance," *Cold Spring Harbor Symp. Quant. Biol.*, **20** (1955), 52–53.

25. T. Nagylaki, "The relation between distant individuals in geographically structured populations," *Math. Biosci.*, **28** (1975), 73–80.
26. W. Feller, *An Introduction to Probability Theory and its Applications*, vol. II, Wiley, New York, 1966.
27. J. B. S. Haldane, "The theory of a cline," *J. Genet.*, **48** (1948), 277–284.
28. M. Slatkin, "Gene flow and selection in a cline," *Genetics*, **75** (1973), 733–756.
29. R. A. Fisher, "The wave of advance of advantageous genes," *Ann. Eugen.*, **7** (1937), 355–369.
30. ———, "Gene frequencies in a cline determined by selection and diffusion," *Biometrics*, **6** (1950), 353–361.
31. T. Nagylaki, "Conditions for the existence of clines," *Genetics*, **80** (1975), 595–615.
32. C. C. Conley, "An application of Wazewski's method to a non-linear boundary value problem which arises in population genetics," *J. Math. Biol.*, **2** (1975), 241–249.
33. L. I. Schiff, *Quantum Mechanics*, 3rd ed., McGraw-Hill, New York, 1968.
34. R. M. May, J. A. Endler and R. E. McMurtrie, "Gene frequency clines in the presence of selection opposed by gene flow," *Amer. Nat.*, **109** (1975), 659–676.
35. W. D. Hanson, "The effects of partial isolation (distance), migration, and different fitness requirements among environmental pockets upon steady state gene frequencies," *Biometrics*, **22** (1966), 453–468.
36. W. H. Fleming, "A selection-migration model in population genetics," *J. Math. Biol.*, **2** (1975), 219–233.
37. M. Slatkin, "Gene flow and selection in a two-locus system," *Genetics*, **81** (1975), 787–802.
38. M. Slatkin and T. Maruyama, "Genetic drift in a cline," *Genetics*, **81** (1975), 209–222.
39. J. Felsenstein, "Genetic drift in clines which are maintained by migration and natural selection," *Genetics*, **81** (1975), 191–207.
40. T. Nagylaki, "The decay of genetic variability in geographically structured populations, II," *Theor. Pop. Biol.*, **10** (1976), 70–82.
41. J. Felsenstein, "A pain in the torus: some difficulties with models of isolation by distance," *Amer. Nat.*, **109** (1975), 359–368.
42. S. Sawyer and J. Felsenstein, "A continuous migration model with stable demography," in preparation (1977).
43. S. Sawyer, "A continuous migration model with stable demography," in preparation (1977).
44. T. Nagylaki, *Selection in One- and Two-Locus Systems*, Springer, Berlin, 1977.

45. S. Sawyer, "Asymptotic properties of the probability of identity in a geographically structured population," *Advances in Appl. Prob.*, **9** (1977), in press.

46. ———, "Results for the stepping stone model for migration in population genetics, *Ann. Prob.*, **4** (1976), 699–728.

47. T. Nagylaki, "Decay of genetic variability in geographically structured populations," *Proc. Nat. Acad. Sci. USA*, **74** (1977), 2523–2525.

48. S. Sawyer, "Rates of consolidation in a selectively neutral migration model," *Ann. Prob.*, **5** (1977), 486–493.

49. ———, "Isotropic random walks in a tree," submitted to *Z. Wahrscheinlichkeitstheorie* (1977).

50. G. Malécot, "Kinship in the birth and death process of a population subdivided in finite panmictic groups," submitted for publication (1977).

51. G. Malécot, "Coefficients of kinship of order n in an infinite regularly subdivided population (one locus case)," submitted for publication (1977).

52. H.-R. Gregorius, "A model for genetic relationship in areally continuous plant populations," *Theor. Pop. Biol.*, **8** (1975), 331–345.

53. P. C. Fife and L. A. Peletier, "Nonlinear diffusion in population genetics," *Arch. Rat. Mech. Analysis*, **64** (1977), 93–109.

54. T. Nagylaki, "Clines with variable migration," *Genetics* **83** (1976), 867–886.

55. ———, "Clines with asymmetric migration," *Genetics* (1978), in press.

56. S. Sawyer, "Results for inhomogeneous selection-migration models," in preparation (1977).

57. T. Nagylaki, "Random genetic drift in a cline," *Proc. Nat. Acad. Sci. USA* (1978), in press.

58. R. A. Fisher, "The advance of advantageous genes," *Ann. Eugen.*, **7** (1937), 355–369.

59. A. Kolmogorov, I. Petrovski, and N. Piskunov, "Étude de l'équation de la diffusion avec croissance de la quantité de matière et son application à un problème biologique," *Bull. Univ. Moskou, Ser. Inter., Sec. A*, **1** No. 6 (1937), 1–37.

60. K. P. Hadeler and F. Rothe, "Travelling fronts in nonlinear diffusion equations," *J. Math. Biol.*, **2** (1975), 251–263.

61. P. C. Fife and J. B. McLeod, "The approach of solutions of nonlinear diffusion equations to travelling wave solutions," *Bull. Amer. Math. Soc.*, **81** (1975), 1075–1078.

62. ———, "The approach of solutions of nonlinear diffusion equations to travelling front solutions," *Arch. Rat. Mech. Analysis*, in press (1977).

63. D. G. Aronson and H. F. Weinberger, "Nonlinear diffusion in population

genetics, combustion, and nerve pulse propagation," in *Partial Differential Equations and Related Topics*, J. Goldstein, ed., Springer, Berlin, 1975.

64. D. G. Aronson and H. F. Weinberger, "Multidimensional nonlinear diffusion arising in population genetics," *Advances in Math.*, in press (1977).

65. H. F. Weinberger, "Asymptotic behavior of a model in population genetics," *Indiana University Seminar in Applied Mathematics*, J. Chadam, ed., Springer, Berlin, in press (1977).

AUTHOR INDEX

This index covers MAA Studies Volume 15 (Studies in Mathematical Biology, Part I, pages 1 to 315) and MAA Studies Volume 16 (Studies in Mathematical Biology, Part II, pages 317–624).

SUBJECT INDEX

This index covers MAA Studies Volume 15 (Studies in Mathematical Biology, Part I, pages 1 to 315) and MAA Studies Volume 16 (Studies in Mathematical Biology, Part II, pages 317–624).